Operators of Fractional Calculus and Their Applications

Operators of Fractional Calculus and Their Applications

Special Issue Editor

Hari Mohan Srivastava

MDPI • Basel • Beijing • Wuhan • Barcelona • Belgrade

MDPI

Special Issue Editor
Hari Mohan Srivastava
University of Victoria
Canada

Editorial Office
MDPI
St. Alban-Anlage 66
Basel, Switzerland

This is a reprint of articles from the Special Issue published online in the open access journal *Mathematics* (ISSN 2227-73900) from 2017 to 2018 (available at: https://www.mdpi.com/journal/mathematics/special_issues/Operators_Fractional_Calculus_Applications)

For citation purposes, cite each article independently as indicated on the article page online and as indicated below:

LastName, A.A.; LastName, B.B.; LastName, C.C. Article Title. *Journal Name* **Year**, *Article Number*, Page Range.

ISBN 978-3-03897-340-9 (Pbk)
ISBN 978-3-03897-341-6 (PDF)

Contents

About the Special Issue Editor

Hari Mohan Srivastava, Prof. Dr., has held the position of Professor Emeritus in the Department of Mathematics and Statistics at the University of Victoria in Canada since 2006, having joined the faculty there in 1969, first as an Associate Professor (1969–1974) and then as a Full Professor (1974–2006). He began his university-level teaching career right after having received his M.Sc. degree in 1959 at the age of 19 years from the University of Allahabad in India. He earned his Ph.D. degree in 1965 while he was a full-time member of the teaching faculty at the Jai Narain Vyas University of Jodhpur in India. He has held numerous visiting research and honorary chair positions at many universities and research institutes in different parts of the world. Having received several D.Sc. (*honoris causa*) degrees as well as honorary memberships and honorary fellowships of many scientific academies and learned societies around the world, he is also actively associated editorially with numerous international scientific research journals. His current research interests include several areas of Pure and Applied Mathematical Sciences, such as Real and Complex Analysis, Fractional Calculus and Its Applications, Integral Equations and Transforms, Higher Transcendental Functions and Their Applications, q-Series and q-Polynomials, Analytic Number Theory, Analytic and Geometric Inequalities, Probability and Statistics, and Inventory Modelling and Optimization. He has published 27 books, monographs, and edited volumes, 30 book (and encyclopedia) chapters, 45 papers in international conference proceedings, and more than 1100 scientific research articles in peer-reviewed international journals, as well as forewords and prefaces to many books and journals, and so on. He is a Clarivate Analytics [Thomson-Reuters] (Web of Science) Highly Cited Researcher. For further details about his other professional achievements and scholarly accomplishments, as well as honors, awards, and distinctions, including the lists of his most recent publications such as journal articles, books, monographs and edited volumes, book chapters, encyclopedia chapters, papers in conference proceedings, forewords to books and journals, et cetera), the interested reader should look into the following regularly updated website.

mathematics Σ

MDPI

Editorial

Operators of Fractional Calculus and Their Applications

Hari Mohan Srivastava [1,2]

1   Department of Mathematics and Statistics, University of Victoria, Victoria, BC V8W 3R4, Canada;
    harimsri@math.uvic.ca
2   Department of Medical Research, China Medical University Hospital, China Medical University,
    Taichung 40402, Taiwan

Received: 4 September 2018; Accepted: 4 September 2018; Published: 5 September 2018

Website: http://www.math.uvic.ca/faculty/harimsri/

This volume contains the successfully invited and accepted submissions (see [1–9]) to a Special Issue of MDPI's journal, *Mathematics* in the subject area of "Operators of Fractional Calculus and Their Applications".

The subject of fractional calculus (that is, calculus of integrals and derivatives of any arbitrary real or complex order) has gained considerable popularity and importance over the past four decades, due, mainly, to its demonstrated applications in numerous diverse and widespread fields of science and engineering. It does indeed provide several potentially useful tools for solving differential, integral, and integro-differential equations, and various other problems involving special functions of Mathematical Physics and Applied Mathematics as well as their extensions and generalizations for one and more variables.

The suggested topics of interest for the call of papers for this Special Issue included, but were not limited to, the following keywords:

- Operators of fractional calculus
- Chaos and fractional dynamics
- Fractional differential
- Fractional differintegral equations
- Fractional integro-differential equations
- Fractional integrals
- Fractional derivatives
- Special Functions of Mathematical Physics and Applied Mathematics
- Identities and inequalities involving fractional integrals

Here, in this Editorial, we briefly describe the status of the Special Issue, as follows:

1. Publications: (8 + 1);
2. Rejections: (16);
3. Article Average Processing Time: 43 days;
4. Article Type: Research Article (8); Review (0); Correction (1)

Authors' geographical distribution:

- Canada (2)
- Korea (2)
- Japan (1)
- India (1)
- Thailand (1)
- Slovakia (1)
- People's Republic of China (1)

- Taiwan (Republic of China) (1)
- Jordan (1)
- USA (1)

The very first work to be devoted exclusively to the subject of fractional calculus, was published in 1974. Ever since then, numerous monographs and books as well as scientific research journals have appeared in the existing literature on the theory and applications of fractional calculus.

Several well-established scientific research journals, published by such publishers as (for example) Elsevier Science Publishers, Hindawi Publishing Corporation, Springer, De Gruyter, MDPI, and others, have published and continue to publish a number of *Special Issues* in many of their journals on recent advances in different aspects of the subject of fractional calculus and its applications. Many widely-attended international conferences, too, continue to be successfully organized and held worldwide ever since the very first one on this subject in USA in the year 1974.

Conflicts of Interest: The author declares no conflict of interest.

References

1. Singh, Y.M.; Khan, M.S.; Kang, S.-M. *F*-Convex contraction via admissible mapping and related fixed point theorems with an application. *Mathematics* **2018**, *6*, 105. [CrossRef]
2. Li, C.-K.; Li, C.P.; Clarkson, K. Several results of fractional differential and integral equations in distribution. *Mathematics* **2018**, *6*, 97. [CrossRef]
3. Li, C.-K.; Clarkson, K. Babenko's approach to Abel's integral equations. *Mathematics* **2018**, *6*, 32. [CrossRef]
4. Morita, T.; Sato, K.-I. Solution of inhomogeneous differential equations with polynomial coefficients in terms of the green's function. *Mathematics* **2017**, *5*, 62. [CrossRef]
5. Fečkan, M.; Wang, J.-R. Mixed order fractional differential equations. *Mathematics* **2017**, *5*, 61. [CrossRef]
6. Yensiri, S.; Skulkhu, R.J. An investigation of radial basis function-finite difference (RBF-FD) method for numerical solution of elliptic partial differential equations. *Mathematics* **2017**, *5*, 54. [CrossRef]
7. Lee, Y.-H. Stability of a monomial functional equation on a restricted domain. *Mathematics* **2017**, *5*, 53. [CrossRef]
8. Thabet, H.; Kendre, S.; Chalishajar, D.K. New analytical technique for solving a system of nonlinear fractional partial differential equations. *Mathematics* **2017**, *5*, 47. [CrossRef]
9. Thabet, H.; Kendre, S.; Chalishajar, D.K. Correction: Thabet, H.; Kendre, S.; Chalishajar, D.K. New analytical technique for solving a system of nonlinear fractional partial differential equations *Mathematics* **2017**, *5*, 47. *Mathematics* **2018**, *6*, 26. [CrossRef]

mathematics

MDPI

Article

New Analytical Technique for Solving a System of Nonlinear Fractional Partial Differential Equations

Hayman Thabet [1], Subhash Kendre [1,*] and Dimplekumar Chalishajar [2]

[1] Department of Mathematics, Savitribai Phule Pune University, Pune 411007, India;
   haymanthabet@gmail.com
[2] Department of Applied Mathematics, Virginia Military Institute, Lexington, VA 24450, USA;
   dipu17370@gmail.com
* Correspondence: sdkendre@yahoo.com

Academic Editor: Hari Mohan Srivastava
Received: 24 August 2017; Accepted: 20 September 2017; Published: 24 September 2017

Abstract: This paper introduces a new analytical technique (NAT) for solving a system of nonlinear fractional partial differential equations (NFPDEs) in full general set. Moreover, the convergence and error analysis of the proposed technique is shown. The approximate solutions for a system of NFPDEs are easily obtained by means of Caputo fractional partial derivatives based on the properties of fractional calculus. However, analytical and numerical traveling wave solutions for some systems of nonlinear wave equations are successfully obtained to confirm the accuracy and efficiency of the proposed technique. Several numerical results are presented in the format of tables and graphs to make a comparison with results previously obtained by other well-known methods

Keywords: system of nonlinear fractional partial differential equations (NFPDEs); systems of nonlinear wave equations; new analytical technique (NAT); existence theorem; error analysis; approximate solution

---

1. Introduction

Over the last few decades, fractional partial differential equations (FPDEs) have been proposed and investigated in many research fields, such as fluid mechanics, the mechanics of materials, biology, plasma physics, finance, and chemistry, and they have played an important role in modeling the so-called anomalous transport phenomena as well as in theory of complex systems, see [1–8]. In study of FPDEs, one should note that finding an analytical or approximate solution is a challenging problem, therefore, accurate methods for finding the solutions of FPDEs are still under investigation. Several analytical and numerical methods for solving FPDEs exist in the literature, for example; the fractional complex transformation [9], homotopy perturbation method [10], a homotopy perturbation technique [11], variational iteration method [12], decomposition method [12], and so on. There are, however, a few solution methods for only traveling wave solutions, for example; the transformed rational function method [13], the multiple exp-function algorithm [14]), and some references cited therein.

The system of NFPDEs have been increasingly used to represent physical and control systems (see for instant, [15–17] and references cited therein). The systems of nonlinear wave equations play an important role in a variety of oceanographic phenomena, for example, in the change in mean sea level due to storm waves, the interaction of waves with steady currents, and the steepening of short gravity waves on the crests of longer waves (see for example, [18–22]). In this paper, two systems of nonlinear wave equations with a fractional order are studied; one is the nonlinear KdV system (see [23,24]) and another one is the system of dispersive long wave equations (see [24–26]).

Some numerical or analytical methods have been investigated for solving a system of NFPDEs, such as an iterative Laplace transform method [27], homotopy analysis method [28], and adaptive observer [29]. Moreover, very few algorithms for the analytical solution of a system of NFPDEs have been suggested, and some of these methods are essentially used for particular types of systems, often just linear ones or even smaller classes. Therefore, it should be noted that most of these methods cannot be generalized to nonlinear cases.

In the present work, we introduce a new analytical technique (NAT) to solve a full general system of NFPDEs of the following form:

$$\begin{cases} \mathcal{D}_t^{q_i} u_i(\bar{x}, t) = f_i(\bar{x}, t) + L_i \bar{u} + N_i \bar{u}, \ m_i - 1 < q_i < m_i \in \mathbb{N}, \ i = 1, 2, \dots, n, \\ \dfrac{\partial^{k_i} u_i}{\partial t^{k_i}}(\bar{x}, 0) = f_{ik_i}(\bar{x}), \ k_i = 0, 1, 2, \dots, m_i - 1, \ i = 1, 2, \dots, n, \end{cases} \tag{1}$$

where L_i and N_i are linear and nonlinear operators, respectively, of $\bar{u} = \bar{u}(\bar{x}, t)$ and its partial derivatives, which might include other fractional partial derivatives of orders less than q_i; $f_i(\bar{x}, t)$ are known analytic functions; and $\mathcal{D}_t^{q_i}$ are the Caputo partial derivatives of fractional orders q_i, where we define $\bar{u} = \bar{u}(\bar{x}, t) = (u_1(\bar{x}, t), u_2(\bar{x}, t), \dots, u_n(\bar{x}, t))$, $\bar{x} = (x_1, x_2, \dots, x_n) \in \mathbb{R}^n$.

The goal of this paper is to demonstrate that a full general system of NFPDEs can be solved easily by using a NAT without any assumption and that it gives good results in analytical and numerical experiments. The rest of the paper is organized in as follows. In Section 2, we present basic definitions and preliminaries which are needed in the sequel. In Section 3, we introduce a NAT for solving a full general system of NFPDEs. Approximate analytical and numerical solutions for the systems of nonlinear wave equations are obtained in Section 4.

2. Basic Definitions and Preliminaries

There are various definitions and properties of fractional integrals and derivatives. In this section, we present modifications of some basic definitions and preliminaries of the fractional calculus theory, which are used in this paper and can be found in [10,30–35].

Definition 1. *A real function $u(x, t)$, $x, t \in \mathbb{R}, t > 0$, is said to be in the space C_μ, $\mu \in \mathbb{R}$ if there exists a real number $p(> \mu)$, such that $u(x, t) = t^p u_1(x, t)$, where $u_1(x, t) \in C(\mathbb{R} \times [0, \infty))$, and it is said to be in the space C_μ^m if and only if $\dfrac{\partial^m u(x,t)}{\partial t^m} \in C_\mu$, $m \in \mathbb{N}$.*

Definition 2. *Let $q \in \mathbb{R} \setminus \mathbb{N}$ and $q \geq 0$. The Riemann–Liouville fractional partial integral denoted by $\mathcal{I}_t^q$ of order q for a function $u(x, t) \in C_\mu$, $\mu > -1$ is defined as:*

$$\begin{cases} \mathcal{I}_t^q u(x, t) = \dfrac{1}{\Gamma(q)} \displaystyle\int_0^t (t - \tau)^{q-1} u(x, \tau) d\tau, \ q, t > 0, \\ \mathcal{I}_t^0 u(x, t) = u(x, t), \quad q = 0, \ t > 0, \end{cases} \tag{2}$$

where Γ is the well-known Gamma function.

Theorem 1. *Let $q_1, q_2 \in \mathbb{R} \setminus \mathbb{N}, q_1, q_2 \geq 0$ and $p > -1$. For a function $u(x, t) \in C_\mu, \mu > -1$, the operator $\mathcal{I}_t^q$ satisfies the following properties:*

$$\begin{cases} \mathcal{I}_t^{q_1} \mathcal{I}_t^{q_2} u(x, t) = \mathcal{I}_t^{q_1 + q_2} u(x, t) \\ \mathcal{I}_t^{q_1} \mathcal{I}_t^{q_2} u(x, t) = \mathcal{I}_t^{q_2} \mathcal{I}_t^{q_1} u(x, t) \\ \mathcal{I}_t^q t^p = \dfrac{\Gamma(p + 1)}{\Gamma(p + q + 1)} t^{p+q} \end{cases} \tag{3}$$

Definition 3. *Let $q, t \in \mathbb{R}, t > 0$ and $u(x,t) \in C_\mu^m$. Then*

$$\begin{cases} \mathcal{D}_t^q u(x,t) = \int_a^t \frac{(t-\tau)^{m-q-1}}{\Gamma(m-q)} \frac{\partial^m u(x,\tau)}{\partial \tau^m} d\tau, & m-1 < q < m \in \mathbb{N}, \\ \mathcal{D}_t^q u(x,t) = \frac{\partial^m u(x,t)}{\partial t^m}, & q = m \in \mathbb{N}, \end{cases} \tag{4}$$

is called the Caputo fractional partial derivative of order q for a function $u(x,t)$.

Theorem 2. *Let $t, q \in \mathbb{R}$, $t > 0$ and $m-1 < q < m \in \mathbb{N}$. Then*

$$\begin{cases} \mathcal{I}_t^q \mathcal{D}_t^q u(x,t) = u(x,t) - \sum_{k=0}^{m-1} \frac{t^k}{k!} \frac{\partial^k u(x,0^+)}{\partial t^k}, \\ \mathcal{D}_t^q \mathcal{I}_t^q u(x,t) = u(x,t). \end{cases} \tag{5}$$

3. NAT for Solving a System of NFPDEs

This section discusses a NAT to solve a system of NFPDEs. This NAT has much more computational power in obtaining piecewise analytical solutions.

To establish our technique, first we need to introduce the following results.

Lemma 1. *For $\bar{u} = \sum_{k=0}^\infty p^k \bar{u}_k$, the linear operator $L_i \bar{u}$ satisfies the following property:*

$$L_i \bar{u} = L_i \sum_{k=0}^\infty p^k \bar{u}_k = \sum_{k=0}^\infty p^k L_i \bar{u}_k, \; i = 1, 2, \ldots, n. \tag{6}$$

Theorem 3. *Let $\bar{u}(\bar{x}, t) = \sum_{k=0}^\infty \bar{u}_k(\bar{x}, t)$, for the parameter λ, we define $\bar{u}_\lambda(\bar{x}, t) = \sum_{k=0}^\infty \lambda^k \bar{u}_k(\bar{x}, t)$, then the nonlinear operator $N_i \bar{u}_\lambda$ satisfies the following property*

$$N_i \bar{u}_\lambda = N_i \sum_{k=0}^\infty \lambda^k \bar{u}_k = \sum_{n=0}^\infty \Big[\frac{1}{n!} \frac{\partial^n}{\partial \lambda^n} [N_i \sum_{k=0}^n \lambda^k \bar{u}_k]_{\lambda=0} \Big] \lambda^n, \; i = 1, 2, \ldots, n. \tag{7}$$

Proof. According to the Maclaurin expansion of $N_i \sum_{k=0}^\infty \lambda^k \bar{u}_k$ with respect to λ, we have

$$N_i \bar{u}_\lambda = N_i \sum_{k=0}^\infty \lambda^k \bar{u}_k = [N_i \sum_{k=0}^\infty \lambda^k \bar{u}_k]_{\lambda=0} + \Big[\frac{\partial}{\partial \lambda} [N_i \sum_{k=0}^\infty \lambda^k \bar{u}_k]_{\lambda=0} \Big] \lambda$$

$$+ \Big[\frac{1}{2!} \frac{\partial^2}{\partial \lambda^2} [N_i \sum_{k=0}^\infty \lambda^k \bar{u}_k]_{\lambda=0} \Big] \lambda^2 + \cdots$$

$$= \sum_{n=0}^\infty \Big[\frac{1}{n!} \frac{\partial^n}{\partial \lambda^n} [N_i \sum_{k=0}^\infty \lambda^k \bar{u}_k]_{\lambda=0} \Big] \lambda^n$$

$$= \sum_{n=0}^\infty \Big[\frac{1}{n!} \frac{\partial^n}{\partial \lambda^n} [N_i (\sum_{k=0}^n \lambda^k \bar{u}_k + \sum_{k=n+1}^\infty \lambda^k \bar{u}_k)]_{\lambda=0} \Big] \lambda^n$$

$$= \sum_{n=0}^\infty \Big[\frac{1}{n!} \frac{\partial^n}{\partial \lambda^n} [N_i \sum_{k=0}^n \lambda^k \bar{u}_k]_{\lambda=0} \Big] \lambda^n, \; i = 1, 2, \ldots, n. \qquad \square$$

Definition 4. *The polynomials $E_{in}(u_{i0}, u_{i1}, \ldots, u_{in})$, for $i = 1, 2, \ldots n$, are defined as*

$$E_{in}(u_{i0}, u_{i1}, \ldots, u_{in}) = \frac{1}{n!} \frac{\partial^n}{\partial \lambda^n} \Big[N_i \sum_{k=0}^n \lambda^k \bar{u}_k \Big]_{\lambda=0}, \; i = 1, 2, \ldots, n. \tag{8}$$

Remark 1. *Let* $E_{in} = E_{in}(u_{i0}, u_{i1}, \ldots, u_{in})$, *by using Theorem* 3 *and Definition* 4, *the nonlinear operators* $N_i \bar{u}_\lambda$ *can be expressed in terms of* E_{in} *as*

$$N_i \bar{u}_\lambda = \sum_{n=0}^{\infty} \lambda^n E_{in}, \ i = 1, 2, \ldots, n. \tag{9}$$

3.1. Existence Theorem

Theorem 4. *Let* $m_i - 1 < q_i < m_i \in \mathbb{N}$ *for* $i = 1, 2, \ldots n$, *and let* $f_i(\bar{x}, t), f_{ik_i}(\bar{x})$ *to be as in* (6), *respectively. Then the system* (1) *admits at least a solution given by*

$$u_i(\bar{x}, t) = \sum_{k_i=0}^{m_i-1} \frac{t^{k_i}}{k_i!} f_{ik_i}(\bar{x}) + f_{it}^{(-q_i)}(\bar{x}, t) + \sum_{k=1}^{\infty} \left[L_{it}^{(-q_i)} \bar{u}_{(k-1)} + E_{i(k-1)t}^{(-q_i)} \right], \ i = 1, 2, \ldots n; \tag{10}$$

where $L_{it}^{(-q_i)} \bar{u}_{(k-1)}$ *and* $E_{i(k-1)t}^{(-q_i)}$ *denote the fractional partial integral of order* q_i *for* $L_{i(k-1)}$ *and* $E_{i(k-1)}$ *respectively with respect to* t.

Proof. Let the solution function $u_i(\bar{x}, t)$ of the system (6) to be as in the following analytical expansion:

$$u_i(\bar{x}, t) = \sum_{k=0}^{\infty} u_{ik}(\bar{x}, t), \ i = 1, 2, \ldots, n. \tag{11}$$

To solve system (1), we consider

$$\mathcal{D}_t^{q_i} u_{i\lambda}(\bar{x}, t) = \lambda \left[f_i(\bar{x}, t) + L_i \bar{u}_\lambda + N_i \bar{u}_\lambda \right], \ i = 1, 2, \ldots, n; \lambda \in [0, 1]. \tag{12}$$

with initial conditions given by

$$\frac{\partial^{k_i} u_{i\lambda}(\bar{x}, 0)}{\partial t^{k_i}} = g_{ik_i}(\bar{x}), \ k_i = 0, 1, 2, \ldots, m_i - 1. \tag{13}$$

Next, we assume that, system (12) has a solution given by

$$u_{i\lambda}(\bar{x}, t) = \sum_{k=0}^{\infty} \lambda^k u_{ik}(\bar{x}, t), \ i = 1, 2, \ldots, n. \tag{14}$$

Performing Riemann-Liouville fractional partial integral of order q_i with respect to t to both sides of system (12) and using Theorem 1, we obtain

$$u_{i\lambda}(\bar{x}, t) = \sum_{k_i=0}^{m_i-1} \frac{t^{k_i}}{k_i!} \frac{\partial^{k_i} u_{i\lambda}(\bar{x}, 0)}{\partial t^{k_i}} + \lambda \mathcal{I}_t^{q_i} \left[f_i(\bar{x}, t) + L_i \bar{u}_\lambda + N_i \bar{u}_\lambda \right], \tag{15}$$

for $i = 1, 2, \ldots, n$. By using the initial condition from the system (1), the system (15) can be rewritten as

$$u_{i\lambda}(\bar{x}, t) = \sum_{k_i=0}^{m_i-1} \frac{t^{k_i}}{k_i!} g_{ik_i}(\bar{x}) + \lambda \left[f_{it}^{(-q_i)}(\bar{x}, t) + \mathcal{I}_t^{q_i} [L_i \bar{u}_\lambda] + \mathcal{I}_t^{q_i} [N_i \bar{u}_\lambda] \right], \tag{16}$$

for $i = 1, 2, \ldots, n$. Inserting (14) into (16), we obtain

$$\sum_{k=0}^{\infty} \lambda^k u_{ik}(\bar{x}, t) = \sum_{k_i=0}^{m_i-1} \frac{t^{k_i}}{k_i!} g_{ik_i}(\bar{x}) + \lambda \left[f_{it}^{(-q_i)}(\bar{x}, t) + \mathcal{I}_t^{q_i} [L_i \sum_{k=0}^{\infty} \lambda^k \bar{u}_k] \right.$$

$$\left. + \mathcal{I}_t^{q_i} [N_i \sum_{k=0}^{\infty} \lambda^k \bar{u}_k] \right], \ i = 1, 2, \ldots, n. \tag{17}$$

By using Lemma 1 and Theorem 3, the system (17) becomes

$$\sum_{k=0}^{\infty} \lambda^k u_{ik}(\bar{x},t) = \sum_{k_i=0}^{m_i-1} \frac{t^{k_i}}{k_i!} g_{ik_i}(\bar{x}) + \lambda f_{it}^{(-q_i)}(\bar{x},t) + \mathcal{I}_t^{q_i} \lambda \sum_{k=0}^{\infty} [L_i \lambda^k \bar{u}_k]$$

$$+ \mathcal{I}_t^{q_i} \lambda \sum_{n=0}^{\infty} \Big[\frac{1}{n!} \frac{\partial^n}{\partial \lambda^n} \big[N_i \sum_{k=0}^{n} \lambda^k \bar{u}_k \big]_{\lambda=0} \Big] \lambda^n, \ i = 1, 2, \ldots, n. \tag{18}$$

Next, we use Definition 4 in the system (18), we obtain

$$\sum_{k=0}^{\infty} \lambda^k u_{ik}(\bar{x},t) = \sum_{k_i=0}^{m_i-1} \frac{t^{k_i}}{k_i!} g_{ik_i}(\bar{x}) + \lambda f_{it}^{(-q_i)}(\bar{x},t) + \mathcal{I}_t^{q_i} \lambda \sum_{k=0}^{\infty} [L_i \lambda^k \bar{u}_k]$$

$$+ \mathcal{I}_t^{q_i} \lambda \sum_{n=0}^{\infty} E_{in} \lambda^n, \ i = 1, 2, \ldots, n. \tag{19}$$

By equating the terms in system (17) with identical powers of λ, we obtain a series of the following systems

$$\begin{cases} u_{i0}(\bar{x},t) = \sum_{k_i=0}^{m_i-1} \frac{t^{k_i}}{k_i!} g_{ik_i}(\bar{x}), \\ u_{i1}(\bar{x},t) = f_{it}^{(-q_i)}(\bar{x},t) + L_{it}^{(-q_i)} \bar{u}_0 + E_{i0t}^{(-q_i)}, \\ u_{i2}(\bar{x},t) = L_{it}^{(-q_i)} \bar{u}_1 + E_{i1t}^{(-q_i)}, \\ \quad \vdots \\ u_{ik}(\bar{x},t) = L_{it}^{(-q_i)} \bar{u}_{(k-1)} + E_{i(k-1)t'}^{(-q_i)} \ k = 2, 3, \ldots, \ i = 1, 2, \ldots, n. \end{cases} \tag{20}$$

Substituting the series (20) in the system (14) gives the solution of the system (12). Now, from the systems (11) and (14), we obtain

$$u_i(\bar{x},t) = \lim_{\lambda \to 1} u_{i\lambda}(\bar{x},t) = u_{i0}(\bar{x},t) + u_{i1}(\bar{x},t) + \sum_{k=2}^{\infty} u_{ik}(\bar{x},t), \ i = 1, 2, \ldots, n. \tag{21}$$

By using the first equations of (21), we see that $\frac{\partial^{k_i} u_i(\bar{x},0)}{\partial t^{k_i}} = \lim_{\lambda \to 1} \frac{\partial^{k_i} u_{i\lambda}(\bar{x},0)}{\partial t^{k_i}}, \ i = 1, 2, \ldots, n,$ which implies that $g_{ik_i}(\bar{x}) = f_{ik_i}(\bar{x}), \ i = 1, 2, \ldots, n.$

Inserting (20) into (21) completes the proof. □

3.2. Convergence and Error Analysis

Theorem 5. *Let B be a Banach space. Then the series solution of the system (20) converges to $S_i \in B$ for $i = 1, 2, \ldots, n$, if there exists γ_i, $0 \le \gamma_i < 1$ such that, $\|u_{in}\| \le \gamma_i \|u_{i(n-1)}\|$ for $\forall n \in \mathbb{N}.$*

Proof. Define the sequences S_{in}, $i = 1, 2, \ldots, n$ of partial sums of the series given by the system (20) as

$$\begin{cases} S_{i0} = u_{i0}(\bar{x},t), \\ S_{i1} = u_{i0}(\bar{x},t) + u_{i1}(\bar{x},t), \\ S_{i2} = u_{i0}(\bar{x},t) + u_{i1}(\bar{x},t) + u_{i2}(\bar{x},t), \\ \quad \vdots \\ S_{in} = u_{i0}(\bar{x},t) + u_{i1}(\bar{x},t) + u_{i2}(\bar{x},t) + \cdots + u_{in}(\bar{x},t), \ i = 1, 2, \ldots, n, \end{cases} \tag{22}$$

and we need to show that $\{S_{in}\}$ are a Cauchy sequences in Banach space B. For this purpose, we consider

$$\|S_{i(n+1)} - S_{in}\| = \|u_{i(n+1)}(\bar{x},t)\| \leq \gamma_i \|u_{in}(\bar{x},t)\| \leq \gamma_i^2 \|u_{i(n-1)}(\bar{x},t)\| \leq \cdots$$
$$\leq \gamma_i^{n+1} \|u_{i0}(\bar{x},t)\|, \ i = 1,2,\ldots,n. \tag{23}$$

For every $n, m \in \mathbb{N}$, $n \geq m$, by using the system (23) and triangle inequality successively, we have,

$$\|S_{in} - S_{im}\| = \|S_{i(m+1)} - S_{im} + S_{i(m+2)} - S_{i(m+1)} + \cdots + S_{in} - S_{i(n-1)}\|$$
$$\leq \|S_{i(m+1)} - S_{im}\| + \|S_{i(m+2)} - S_{i(m+1)}\| + \cdots + \|S_{in} - S_{i(n-1)}\|$$
$$\leq \gamma_i^{m+1} \|u_{i0}(\bar{x},t)\| + \gamma_i^{m+2} \|u_{i0}(\bar{x},t)\| + \cdots + \gamma_i^n \|u_{i0}(\bar{x},t)\|$$
$$= \gamma_i^{m+1}(1 + \gamma_i + \cdots + \gamma_i^{n-m-1}) \|u_{i0}(\bar{x},t)\|$$
$$\leq \gamma_i^{m+1} \left(\frac{1 - \gamma^{n-m}}{1 - \gamma_i} \right) \|u_{i0}(\bar{x},t)\|. \tag{24}$$

Since $0 < \gamma_i < 1$, so $1 - \gamma_i^{n-m} \leq 1$ then

$$\|S_{in} - S_{im}\| \leq \frac{\gamma_i^{m+1}}{1 - \gamma_i} \|u_{i0}(\bar{x},t)\|. \tag{25}$$

Since $u_{i0}(\bar{x},t)$ is bounded, then

$$\lim_{n,m \to \infty} \|S_{in} - S_{im}\| = 0, \ i = 1,2,\ldots,n. \tag{26}$$

Therefore, the sequences $\{S_{in}\}$ are Cauchy sequences in the Banach space B, so the series solution defined in the system (21) converges. This completes the proof. $\square$

Theorem 6. *The maximum absolute truncation error of the series solution* (11) *of the nonlinear fractional partial differential system* (1) *is estimated to be*

$$\sup_{(\bar{x},t) \in \Omega} \left| u_i(\bar{x},t) - \sum_{k=0}^{m} u_{ik}(\bar{x},t) \right| \leq \frac{\gamma_i^{m+1}}{1 - \gamma_i} \sup_{(\bar{x},t) \in \Omega} |u_{i0}(\bar{x},t)|, \ i = 1,2,\ldots,n, \tag{27}$$

where the region $\Omega \subset \mathbb{R}^{n+1}$.

Proof. From Theorem 5, we have

$$\|S_{in} - S_{im}\| \leq \frac{\gamma_i^{m+1}}{1 - \gamma_i} \sup_{(\bar{x},t) \in \Omega} |u_{i0}(\bar{x},t)|, \ i = 1,2,\ldots,n. \tag{28}$$

But we assume that $S_{in} = \sum_{k=0}^{n} u_{ik}(\bar{x},t)$ for $i = 1,2,\ldots,n$, and since $n \to \infty$, we obtain $S_{in} \to u_i(\bar{x},t)$, so the system (28) can be rewritten as

$$\|u_i(\bar{x},t) - S_{im}\| = \|u_i(\bar{x},t) - \sum_{k=0}^{m} u_{ik}(\bar{x},t)\|$$
$$\leq \frac{\gamma_i^{m+1}}{1 - \gamma_i} \sup_{(\bar{x},t) \in \Omega} |u_{i0}(\bar{x},t)|, \ i = 1,2,\ldots,n. \tag{29}$$

So, the maximum absolute truncation error in the region Ω is

$$\sup_{(\bar{x},t)\in\Omega}\left|u_i(\bar{x},t)-\sum_{k=0}^{m}u_{ik}(\bar{x},t)\right| \leq \frac{\gamma_i^{m+1}}{1-\gamma_i}\sup_{(\bar{x},t)\in\Omega}|u_{i0}(\bar{x},t)|, \ i=1,2,\ldots,n. \tag{30}$$

and this completes the proof. $\square$

4. Applications to the Systems of Nonlinear Wave Equations

In this section, we present examples of some systems of nonlinear wave equations. These examples are chosen because their closed form solutions are available, or they have been solved previously by some other well-known methods.

Example 1. *Consider the nonlinear KdV system of time-fractional order of the form* [24]

$$D_t^q u = -\alpha u_{xxx} - 6\alpha u u_x + 6v v_x, \ D_t^q v = -\alpha v_{xxx} - 3\alpha u v_x, \tag{31}$$

for $0 < q < 1$, subject to the initial conditions

$$u(x,0) = \beta^2 \mathrm{sech}^2(\frac{\gamma}{2}+\frac{\beta x}{2}), \ v(x,0) = \sqrt{\frac{\alpha}{2}}\beta^2 \mathrm{sech}^2(\frac{\gamma}{2}+\frac{\beta x}{2}). \tag{32}$$

For $q = 1$, the exact solitary wave solutions of the KdV system (31) *is given by*

$$\begin{cases} u(x,t) = \beta^2 \mathrm{sech}^2(\frac{1}{2}[\gamma - \alpha\beta^3 t + \beta x]), \\ v(x,t) = \sqrt{\frac{\alpha}{2}}\beta^2 \mathrm{sech}^2(\frac{1}{2}[\gamma - \alpha\beta^3 t + \beta x]), \end{cases} \tag{33}$$

where the constant α is a wave velocity and β, γ are arbitrary constants.

To solve the system (31), we compare (31) with the system (1), we obtain

$$D_t^q u = -\alpha u_{xxx} + N_1(u,v), \ D_t^q v = -\alpha v_{xxx} + N_2(u,v), \tag{34}$$

where we assume $N_1(u,v) = 6vv_x - 6\alpha u u_x$ and $N_2(u,v) = -3\alpha u v_x$.

Next, we assume the system (31) has a solution given by

$$u(x,t) = \sum_{k=0}^{\infty} u_k(x,t), \ v(x,t) = \sum_{k=0}^{\infty} v_k(x,t). \tag{35}$$

To obtain the approximate solution of the system (31), we consider the following system.

$$\mathcal{D}_t^q u_\lambda = \lambda\big[-\alpha u_{\lambda xxx} + N_1(u_\lambda, v_\lambda)\big], \ \mathcal{D}_t^q v_\lambda = \lambda\big[-\alpha v_{\lambda xxx} + N_2(u_\lambda, v_\lambda)\big], \tag{36}$$

subject to the initial conditions given by

$$u_\lambda(x,0) = g_1(x), \ v_\lambda(x,0) = g_2(x), \tag{37}$$

and we assume that the system (36) has a solution of the form

$$u_\lambda(x,t) = \sum_{k=0}^{\infty} \lambda^k u_k(x,t), \ v_\lambda(x,t) = \sum_{k=0}^{\infty} \lambda^k v_k(x,t). \tag{38}$$

By operating Riemann-Liouville fractional partial integral of order q with respect to t for both sides of the system (36) and by using Theorem 2 and the system (37), we obtain

$$\begin{cases} u_\lambda = g_1(x) + \lambda \mathcal{I}_t^q \big[-\alpha u_{\lambda xxx} + N_1(u_\lambda, v_\lambda) \big], \\ v_\lambda = g_2(x) + \lambda \mathcal{I}_t^q \big[-\alpha v_{\lambda xxx} + N_2(u_\lambda, v_\lambda) \big] \end{cases} \tag{39}$$

By using Remark 1 and system (38), in the system (39), we obtain

$$\begin{cases} \sum_{k=0}^{\infty} \lambda^k u_k = g_1(x) + \lambda \mathcal{I}_t^q \big[-\alpha \sum_{k=0}^{\infty} \lambda^k u_{kxxx} + \sum_{n=0}^{\infty} \lambda^n E_{1n} \big], \\ \sum_{k=0}^{\infty} \lambda^k v_k = g_2(x) + \lambda \mathcal{I}_t^q \big[-\alpha \sum_{k=0}^{\infty} \lambda^k v_{kxxx} + \sum_{n=0}^{\infty} \lambda^n E_{2n} \big]. \end{cases} \tag{40}$$

By equating the terms in the system (40) with identical powers of λ, we obtain a series of the following systems.

$$\begin{cases} u_0 = g_1(x), \ v_0 = g_2(x), \\ u_1 = \mathcal{I}_t^q \big[-\alpha u_{0xxx} + E_{10} \big], \ v_1 = \mathcal{I}_t^q \big[-\alpha v_{0xxx} + E_{20} \big], \\ u_2 = \mathcal{I}_t^q \big[-\alpha u_{1xxx} + E_{11} \big], \ v_2 = \mathcal{I}_t^q \big[-\alpha v_{1xxx} + E_{21} \big], \\ \vdots \\ u_k = \mathcal{I}_t^q \big[-\alpha u_{(k-1)xxx} + E_{1(k-1)} \big], \ v_k = \mathcal{I}_t^q \big[-\alpha v_{(k-1)xxx} + E_{2(k-1)} \big], \end{cases} \tag{41}$$

for $k = 1, 2, \ldots$, where $E_{1(k-1)}, E_{1(k-1)}$ can be obtain by using Definition 4.

By using the systems (35) and (38), we can set

$$u(x,t) = \lim_{\lambda \to 1} u_\lambda(x,t) = \sum_{k=0}^{\infty} u_k(x,t), \ v(x,t) = \lim_{\lambda \to 1} v_\lambda(x,t) = \sum_{k=0}^{\infty} v_k(x,t). \tag{42}$$

By using the first equations of (42), we have $u(x,0) = \lim_{\lambda \to 1} u_\lambda(x,0)$, $v(x,0) = \lim_{\lambda \to 1} v_\lambda(x,0)$, which implies that $g_1(x) = u(x,0)$ and $g_2(x) = v(x,0)$. Consequently, by using (41) and Definition 4, with the help of Mathematica software, the first few components of the solution for the system (31) are derived as follows.

$$u_0(x,t) = \beta^2 \text{sech}^2(\frac{\gamma}{2} + \frac{\beta x}{2}), \ v_0(x,t) = \sqrt{\frac{\alpha}{2}} \beta^2 \text{sech}^2(\frac{\gamma}{2} + \frac{\beta x}{2}),$$

$$u_1(x,t) = \frac{\alpha \beta^5}{\Gamma(q+1)} \tanh(\frac{\gamma}{2} + \frac{\beta x}{2}) \text{sech}^2(\frac{\gamma}{2} + \frac{\beta x}{2}) t^q,$$

$$v_1(x,t) = \frac{\alpha^{3/2} \beta^5}{\sqrt{2} \Gamma(q+1)} \tanh(\gamma 2 + \frac{\beta x}{2}) \text{sech}^2(\frac{\gamma}{2} + \frac{\beta x}{2}) t^q,$$

$$u_2(x,t) = \frac{\alpha^2 \beta^8}{2 \Gamma(2q+1)} [\cosh(\gamma + \beta x) - 2] \text{sech}^4(\frac{\gamma}{2} + \frac{\beta x}{2})) t^{2q},$$

$$v_2(x,t) = \frac{\alpha^{5/2} \beta^8}{2\sqrt{2} \Gamma(2q+1)} (\cosh(\gamma + \beta x) - 2) \text{sech}^4(\frac{\gamma}{2} + \frac{\beta x}{2}) t^{2q},$$

$$u_3(x,t) = \frac{\alpha^3 \beta^{11}}{8 \Gamma(q+1)^2 \Gamma(3q+1)} \big[\Gamma(q+1)^2 [-32 \cosh(\gamma + \beta x) + \cosh(2[\gamma + \beta x])$$

$$+ 39] + 12 \Gamma(2q+1) [\cosh(\gamma + \beta x) - 2] \big] \tanh(\frac{\gamma}{2} + \frac{\beta x}{2}) \text{sech}^6(\frac{\gamma}{2} + \frac{\beta x}{2}) t^{3q},$$

$$v_3(x,t) = \frac{\alpha^{7/2}\beta^{11}}{8\sqrt{2}\Gamma(q+1)^2\Gamma(3q+1)}[\Gamma(q+1)^2[-32\cosh(\gamma+\beta x)+\cosh(2(\gamma+\beta x))$$

$$+39]+12\Gamma(2q+1)[\cosh(\gamma+\beta x)-2]]\tanh(\frac{\gamma}{2}+\frac{\beta x}{2})\text{sech}^6(\frac{\gamma}{2}+\frac{\beta x}{2})t^{3q},$$

$$\vdots$$

and so on.

Hence the third-order term approximate solution for the system (31) is given by

$$u(x,t) = \beta^2\text{sech}^2(\frac{\gamma}{2}+\frac{\beta x}{2}) + \frac{\alpha\beta^5}{\Gamma(q+1)}\tanh(\frac{\gamma}{2}+\frac{\beta x}{2})\text{sech}^2(\frac{\gamma}{2}+\frac{\beta x}{2})t^q$$

$$+\frac{\alpha^2\beta^8}{2\Gamma(2q+1)}[\cosh(\gamma+\beta x)-2]\text{sech}^4(\frac{\gamma}{2}+\frac{\beta x}{2}))t^{2q}$$

$$+\frac{\alpha^3\beta^{11}}{8\Gamma(q+1)^2\Gamma(3q+1)}[\Gamma(q+1)^2[-32\cosh(\gamma+\beta x)+\cosh(2[\gamma+\beta x])$$

$$+39]+12\Gamma(2q+1)[\cosh(\gamma+\beta x)-2]]\tanh(\frac{\gamma}{2}+\frac{\beta x}{2})\text{sech}^6(\frac{\gamma}{2}+\frac{\beta x}{2})t^{3q},$$

$$v(x,t) = \sqrt{\frac{\alpha}{2}}\beta^2\text{sech}^2(\frac{\gamma}{2}+\frac{\beta x}{2}) + \frac{\alpha^{3/2}\beta^5}{\sqrt{2}\Gamma(q+1)}\tanh(\gamma 2+\frac{\beta x}{2})\text{sech}^2(\frac{\gamma}{2}+\frac{\beta x}{2})t^q$$

$$+\frac{\alpha^{5/2}\beta^8}{2\sqrt{2}\Gamma(2q+1)}(\cosh(\gamma+\beta x)-2)\text{sech}^4(\frac{\gamma}{2}+\frac{\beta x}{2})t^{2q}$$

$$+\frac{\alpha^{7/2}\beta^{11}}{8\sqrt{2}\Gamma(q+1)^2\Gamma(3q+1)}[\Gamma(q+1)^2[-32\cosh(\gamma+\beta x)+\cosh(2(\gamma+\beta x))$$

$$+39]+12\Gamma(2q+1)[\cosh(\gamma+\beta x)-2]]\tanh(\frac{\gamma}{2}+\frac{\beta x}{2})\text{sech}^6(\frac{\gamma}{2}+\frac{\beta x}{2})t^{3q}.$$

In Table 1, the numerical values of the approximate and exact solutions for Example 1 show the accuracy and efficiency of our technique at different values of x, t. The absolute error is listed for different values of x, t. In Figure 1a, we consider fixed values $\alpha = \beta = 0.5, \gamma = 1$ and fixed order $q = 1$ for piecewise approximation values of x, t in the domain $-20 \le x \le 20$ and $0.20 \le t \le 1$. In Figure 1b, we plot the exact solution with fixed values $\alpha = \beta = 0.5$ and $\gamma = 1$ in the domain $-20 \le x \le 20$ and $0.20 \le t \le 1$.

Table 1. Numerical values when $q = 0.5, 1$ and $\alpha = \beta = 0.5, \gamma = 1$ for Example 1.

x	t	$q=0.5$		$q=1$		$\alpha=\beta=0.5, \gamma=1$		Absolute Error	
		u_{NAT}	v_{NAT}	u_{NAT}	v_{NAT}	u_{EX}	v_{EX}	$\|u_{EX}-u_{NAT}\|$	$\|v_{EX}-v_{NAT}\|$
	0.20	0.0171378	0.0085689	0.0174511	0.0087256	0.0174511	0.0087256	9.11712×10^{-12}	4.55856×10^{-12}
-10	0.40	0.0169274	0.0084637	0.0172419	0.0086210	0.0172419	0.0086210	1.45834×10^{-10}	7.29172×10^{-11}
	0.60	0.0167686	0.0083843	0.0170352	0.0085176	0.0170352	0.0085176	7.38075×10^{-10}	3.69037×10^{-10}
	0.20	0.1994480	0.0997242	0.1977450	0.0988724	0.1977450	0.0988724	5.11989×10^{-11}	2.55994×10^{-11}
0	0.40	0.2006050	0.1003020	0.1988720	0.0994360	0.1988720	0.0994360	8.07505×10^{-10}	4.03753×10^{-10}
	0.60	0.20148400	0.1007420	0.1999930	0.0999966	0.1999930	0.0999966	4.02841×10^{-9}	2.01421×10^{-9}
	0.20	0.0000172	8.62×10^{-6}	0.0000169	8.46×10^{-6}	0.0000169	8.46×10^{-6}	1.70233×10^{-14}	8.51164×10^{-15}
20	0.40	0.0000175	8.74×10^{-6}	0.0000171	8.56×10^{-6}	0.0000171	8.56×10^{-6}	2.73056×10^{-13}	1.36528×10^{-13}
	0.60	0.0000177	8.83×10^{-6}	0.0000173	8.67×10^{-6}	0.0000173	8.67×10^{-6}	1.38582×10^{-12}	6.92909×10^{-13}

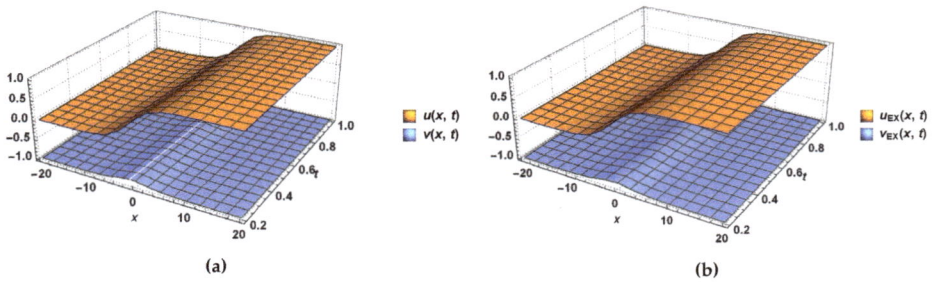

Figure 1. (a) The graph for the approximate solution of Example 2 for $\alpha = \beta = 0.5$ and $q_1 = q_2 = 1$; (b) The graph for the exact solution of Example 2 for $\alpha = \beta = 0.5$.

Example 2. *Consider the nonlinear dispersive long wave system of time fractional order* [24–26]

$$D_t^{q_1} u = -v_x - \frac{1}{2}(u^2)_x, \; D_t^{q_2} v = -(u + u_{xx} + uv)_x, \tag{43}$$

for $0 < q_1, q_1 < 1$, with initial condition given by

$$u(x,0) = \alpha[\tanh(\frac{1}{2}[\beta + \alpha x]) + 1], \; v(x,0) = -1 + \frac{1}{2}\alpha^2 sech^2(\frac{1}{2}[\beta + \alpha x]). \tag{44}$$

For $q_1 = q_2 = 1$, the system (43) has the following exact solitary wave solutions:

$$u(x,t) = \alpha[\tanh(\frac{1}{2}[\beta + \alpha x - \alpha^2 t]) + 1], \; v(x,t) = -1 + \frac{1}{2}\alpha^2 sech^2(\frac{1}{2}[\beta + \alpha x - \alpha^2 t]), \tag{45}$$

where α, β are arbitrary constants.

By comparing the system (43) with the system (1), the system (43) can be rewritten as

$$D_t^{q_1} u = -v_x + N_1(u,v), \; D_t^{q_2} v = -u_x - u_{xxx} + N_2(u,v), \tag{46}$$

where $N_1(u,v) = -uu_x$ and $N_2(u,v) = -(uv_x + vu_x)$. To solve the system (46) by NAT discussed in Section 3, we assume that the system (46) has a solution given by

$$u(x,t) = \sum_{k=0}^{\infty} u_k(x,t), \; v(x,t) = \sum_{k=0}^{\infty} v_k(x,t). \tag{47}$$

Forgetting the approximate solution of the system (43), we consider the following system.

$$D_t^{q_1} u_\lambda = \lambda[-v_{x\lambda} + N_1(u_\lambda, v_\lambda)], \; D_t^{q_2} v_\lambda = \lambda[-u_{\lambda x} - u_{\lambda xxx} + N_2(u_\lambda, v_\lambda)], \tag{48}$$

subject to the initial conditions given by

$$u_\lambda(x,0) = g_1(x), \; v_\lambda(x,0) = g_2(x). \tag{49}$$

Assume that the system (48) has a solution given by

$$u_\lambda(x,t) = \sum_{k=0}^{\infty} \lambda^k u_k(x,t), \; v_\lambda(x,t) = \sum_{k=0}^{\infty} \lambda^k v_k(x,t). \tag{50}$$

By using Theorem 2, we take Riemann-Liouville fractional partial integrals of order q_1 and q_2 with respect to t for both sides of the system (48) and using (47), we obtain

$$\begin{cases} u_\lambda = g_1(x) + \lambda \mathcal{I}_t^{q_1}\left[-v_{x\lambda} + N_1(u_\lambda, v_\lambda)\right], \\ v_\lambda = g_2(x) + \lambda \mathcal{I}_t^{q_2}\left[-u_{x\lambda} - u_{xxx\lambda} + N_2(u_\lambda, v_\lambda)\right]. \end{cases} \tag{51}$$

Next, we use Theorem 1. The system (51) can be rewritten as

$$\begin{cases} \displaystyle\sum_{k=0}^{\infty} \lambda^k u_k = g_1(x) + \lambda \mathcal{I}_t^{q_1}\left[-\sum_{k=0}^{\infty} \lambda^k v_{xk} + \sum_{n=0}^{\infty} \lambda^n E_{1n}\right], \\ \displaystyle\sum_{k=0}^{\infty} \lambda^k v_k = g_2(x) + \lambda \mathcal{I}_t^{q_2}\left[-\sum_{k=0}^{\infty} \lambda^k u_{kx} - \sum_{k=0}^{\infty} \lambda^k u_{kxxx} + \sum_{n=0}^{\infty} \lambda^n E_{2n}\right]. \end{cases} \tag{52}$$

By equating the terms in the system (52) with identical powers of λ, we obtain a series of the following systems.

$$\begin{cases} u_0 = g_1(x), \; v_0 = g_2(x), \\ u_1 = \mathcal{I}_t^{q_1}\left[-v_{0x} + E_{10}\right], \; v_1 = \mathcal{I}_t^{q_2}\left[-u_{0x} - u_{0xxx} + E_{20}\right], \\ u_2 = \mathcal{I}_t^{q_1}\left[-v_{1x} + E_{11}\right], \; v_2 = \mathcal{I}_t^{q_2}\left[-u_{1x} - u_{1xxx} + E_{21}\right], \\ \vdots \\ u_k = \mathcal{I}_t^{q_1}\left[-v_{(k-1)x} + E_{1(k-1)}\right], \; v_k = \mathcal{I}_t^{q_2}\left[-u_{(k-1)x} - u_{(k-1)xxx} + E_{2(k-1)}\right], \end{cases} \tag{53}$$

for $k = 1, 2, \ldots$, where $E_{1(k-1)}, E_{1(k-1)}$ can be obtain by using Theorem 4.

From the systems (47) and (50), we have

$$u(x,t) = \lim_{\lambda \to 1} u_\lambda(x,t) = \sum_{k=0}^{\infty} u_k(x,t), \; v(x,t) = \lim_{\lambda \to 1} v_\lambda(x,t) = \sum_{k=0}^{\infty} v_k(x,t). \tag{54}$$

By using the first equations of (54), we have $u(x,0) = \lim_{\lambda \to 1} u_\lambda(x,0)$, $v(x,0) = \lim_{\lambda \to 1} v_\lambda(x,0)$, which implies that $g_1(x) = u(x,0)$ and $g_2(x) = v(x,0)$. Consequently, by using (53) and Definition 4 by the help of Mathematica software, the first few components of the solution for the system (43) are derived as follows.

$$u_0(x,t) = \alpha\left[\tanh\left(\frac{1}{2}[\beta + \alpha x]\right) + 1\right], \; v_0(x,t) = -1 + \frac{1}{2}\alpha^2 \text{sech}^2\left(\frac{1}{2}[\beta + \alpha x]\right),$$

$$u_1(x,t) = -\frac{\alpha^3}{2q_1\Gamma(q_1)}\text{sech}^2\left(\frac{1}{2}[\beta + \alpha x]\right)t^{q_1},$$

$$v_1(x,t) = \frac{4\alpha^4}{q_2\Gamma(q_2)}\sinh^4\left(\frac{1}{2}[\beta + \alpha x]\right)\text{csch}^3(\beta + \alpha x)t^{q_2},$$

$$u_2(x,t) = \frac{1}{4}\alpha^5\text{sech}^4\left(\frac{1}{2}[\beta + \alpha x]\right)\left[\frac{[\cosh(\beta + \alpha x) - 2]t^{q_2}}{\Gamma(q_1 + q_2 + 1)}\right.$$
$$\left. - \frac{[\sinh(\beta + \alpha x) + \cosh(\beta + \alpha x) - 2]t^{q_1}}{\Gamma(2q_1 + 1)}\right]t^{q_1},$$

$$v_2(x,t) = \frac{\alpha^6\text{sech}^5\left(\frac{1}{2}[\beta + \alpha x]\right)}{8\Gamma(q_1 + q_2 + 1)\Gamma(2q_2 + 1)}t^{q_2}\left[t^{q_1}\Gamma(2q_2 + 1)\left[7\sinh\left(\frac{1}{2}[\beta + \alpha x]\right)\right.\right.$$
$$\left. - \sinh\left(\frac{3}{2}[\beta + \alpha x]\right)\right] - \Gamma(q_1 + q_2 + 1)\left[7\sinh\left(\frac{1}{2}[\beta + \alpha x]\right)\right.$$
$$\left.\left. - \sinh\left(\frac{3}{2}[\beta + \alpha x]\right) + 3\cosh\left(\frac{1}{2}[\beta + \alpha x]\right) - \cosh(32[\beta + \alpha x])\right]t^{q_2}\right],$$

$$u_3(x,t) = \frac{1}{16}\alpha^7 t^{q_1}\left[\frac{64t^{2q_1}e^{2(\beta+\alpha x)}}{\Gamma(3q_1+1)\left(e^{\beta+\alpha x}+1\right)^6}\left[\frac{4^{q_1}\Gamma\left[q_1+\frac{1}{2}\right]\left[e^{2(\beta+\alpha x)}-1\right]}{\sqrt{\pi}\Gamma(q_1+1)}\right.\right.$$

$$-2e^{\beta+\alpha x}\left[e^{\beta+\alpha x}[e^{\beta+\alpha x}-8]+6\right]\Big]+t^{q_2}\text{sech}^6(\tfrac{1}{2}[\beta+\alpha x])$$

$$\times\left[\frac{t^{q_2}e^{\alpha(-x)-\beta}\left[e^{\beta+\alpha x}[e^{\beta+\alpha x}[e^{\beta+\alpha x}-14]+21]-4\right]}{\Gamma(q_1+2q_2+1)}\right.$$

$$+\frac{t^{q_1}\left[-10\sinh(\beta+\alpha x)+\sinh(2[\beta+\alpha x])+4\cosh(\beta+\alpha x)-6\right]}{\Gamma(2q_1+q_2+1)}\Bigg]\Bigg],$$

$$v_3(x,t) = \frac{2\alpha^8 t^{q_2}e^{\beta+\alpha x}}{\left[e^{\beta+\alpha x}+1\right]^7}\left[t^{q_2}\left[\frac{4t^{q_2}e^{2(\beta+\alpha x)}\left[e^{\beta+\alpha x}\left(e^{\beta+\alpha x}-13\right)\left(e^{\beta+\alpha x}-2\right)-6\right]}{\Gamma(3q_2+1)}\right.\right.$$

$$+\left[-\frac{4\Gamma(q_1+q_2+1)e^{\beta+\alpha x}\left[e^{\beta+\alpha x}+1\right]\left[e^{\beta+\alpha x}\left(e^{\beta+\alpha x}-3\right)+1\right]}{\Gamma(q_1+1)\Gamma(q_2+1)\Gamma(q_1+2q_2+1)}\right.$$

$$+\frac{e^{\beta+\alpha x}\left[49-e^{\beta+\alpha x}\left[e^{\beta+\alpha x}\left[e^{\beta+\alpha x}\left(e^{\beta+\alpha x}+15\right)-172\right]+220\right]\right]}{\Gamma(q_1+2q_2+1)}-1\Big]t^{q_1}\Bigg]$$

$$-\frac{2t^{2q_1}e^{\beta+\alpha x}\left[e^{\beta+\alpha x}\left(e^{\beta+\alpha x}\left[e^{\beta+\alpha x}\left(e^{\beta+\alpha x}-37\right)+151\right]-119\right)+16\right]}{\Gamma(2q_1+q_2+1)}\Bigg],$$

$$\vdots$$

and so on.

Hence the third-order term approximate solution for the system (43) is given by

$$u(x,t) = \alpha[\tanh(\tfrac{1}{2}[\beta+\alpha x])+1]-\frac{\alpha^3}{2q_1\Gamma(q_1)}\text{sech}^2(\tfrac{1}{2}[\beta+\alpha x])t^{q_1}$$

$$+\frac{1}{4}\alpha^5\text{sech}^4(\tfrac{1}{2}[\beta+\alpha x])\left[\frac{[\cosh(\beta+\alpha x)-2]t^{q_2}}{\Gamma(q_1+q_2+1)}\right.$$

$$-\frac{[\sinh(\beta+\alpha x)+\cosh(\beta+\alpha x)-2]t^{q_1}}{\Gamma(2q_1+1)}\Big]t^{q_1}$$

$$+\frac{1}{16}\alpha^7 t^{q_1}\left[\frac{64t^{2q_1}e^{2(\beta+\alpha x)}}{\Gamma(3q_1+1)\left(e^{\beta+\alpha x}+1\right)^6}\left[\frac{4^{q_1}\Gamma\left[q_1+\frac{1}{2}\right]\left[e^{2(\beta+\alpha x)}-1\right]}{\sqrt{\pi}\Gamma(q_1+1)}\right.\right.$$

$$-2e^{\beta+\alpha x}\left[e^{\beta+\alpha x}[e^{\beta+\alpha x}-8]+6\right]\Big]+t^{q_2}\text{sech}^6(\tfrac{1}{2}[\beta+\alpha x])$$

$$\times\left[\frac{t^{q_2}e^{\alpha(-x)-\beta}\left[e^{\beta+\alpha x}[e^{\beta+\alpha x}[e^{\beta+\alpha x}-14]+21]-4\right]}{\Gamma(q_1+2q_2+1)}\right.$$

$$+\frac{t^{q_1}(-10\sinh(\beta+\alpha x)+\sinh(2(\beta+\alpha x))+4\cosh(\beta+\alpha x)-6)}{\Gamma(2q_1+q_2+1)}\Bigg]\Bigg],$$

$$v(x,t) = -1+\frac{1}{2}\alpha^2\text{sech}^2(\tfrac{1}{2}[\beta+\alpha x])+\frac{4\alpha^4}{q_2\Gamma(q_2)}\sinh^4(\tfrac{1}{2}[\beta+\alpha x])\text{csch}^3(\beta+\alpha x)t^{q_2}$$

$$+\frac{\alpha^6\text{sech}^5(\tfrac{1}{2}[\beta+\alpha x])}{8\Gamma(q_1+q_2+1)\Gamma(2q_2+1)}t^{q_2}\left[t^{q_1}\Gamma(2q_2+1)\left[7\sinh(\tfrac{1}{2}[\beta+\alpha x])\right.\right.$$

$$-\sinh(\tfrac{3}{2}[\beta+\alpha x])\Big]-\Gamma(q_1+q_2+1)\left[7\sinh(\tfrac{1}{2}[\beta+\alpha x])\right.$$

$$-\sinh(\tfrac{3}{2}[\beta+\alpha x])+3\cosh(\tfrac{1}{2}[\beta+\alpha x])-\cosh(32[\beta+\alpha x])\Big]t^{q_2}\Bigg]$$

$$+ \frac{2\alpha^8 t^{q_2} e^{\beta+\alpha x}}{[e^{\beta+\alpha x}+1]^7} \left[t^{q_2} \left[\frac{4t^{q_2} e^{2(\beta+\alpha x)} \left[e^{\beta+\alpha x} [e^{\beta+\alpha x} - 13] [e^{\beta+\alpha x} - 2] - 6 \right]}{\Gamma(3q_2+1)} \right. \right.$$

$$+ \left[- \frac{4\Gamma(q_1+q_2+1) e^{\beta+\alpha x} [e^{\beta+\alpha x}+1] \left[e^{\beta+\alpha x} [e^{\beta+\alpha x} - 3] + 1 \right]}{\Gamma(q_1+1)\Gamma(q_2+1)\Gamma(q_1+2q_2+1)} \right.$$

$$+ \left. \left. \frac{e^{\beta+\alpha x} \left[49 - e^{\beta+\alpha x} [e^{\beta+\alpha x} [e^{\beta+\alpha x} [e^{\beta+\alpha x}+15] - 172] + 220] \right]}{\Gamma(q_1+2q_2+1)} - 1 \right] t^{q_1} \right]$$

$$\left. - \frac{2t^{2q_1} e^{\beta+\alpha x} \left[e^{\beta+\alpha x} [e^{\beta+\alpha x} [e^{\beta+\alpha x} [e^{\beta+\alpha x} - 37] + 151] - 119] + 16 \right]}{\Gamma(2q_1+q_2+1)} \right].$$

In Table 2, we obtain similar approximation values as in the exact solution for Example 2 at different values of x, t. The absolute error is listed against different values of x, t. In Figure 2a, we plot the approximate solution at fixed values of the constants $\alpha = \beta = 0.5$ and fixed order $q = 1$. In Figure 2b, we plot the exact solution with fixed values of the constants $\alpha = \beta = 0.5$. The graphs of the approximate and exact solutions are drawing in the domain $-20 \le x \le 20$ and $0.20 \le t \le 1$.

Table 2. Numerical values when $q_1 = q_2 = 0.5, 1$ and $\alpha = \beta = 0.5$ for Example 2.

x	t	$q_1 = 0.5$	$q_2 = 0.5$	$q_1 = 1$	$q_2 = 1$	$\alpha = 0.5$	$\beta = 0.5$	Absolute Error	
		u_{NAT}	v_{NAT}	u_{NAT}	v_{NAT}	u_{EX}	v_{EX}	$\lvert u_{EX} - u_{NAT} \rvert$	$\lvert v_{EX} - v_{NAT} \rvert$
	0.20	0.0104859	-0.9952150	0.0104567	-0.9948260	0.0104567	-0.9948260	2.38731×10^{-9}	9.68267×10^{-10}
-10	0.40	0.0100181	-0.9954506	0.0099518	-0.9950740	0.0099518	-0.9950740	3.78897×10^{-8}	1.54049×10^{-8}
	0.60	0.0095700	-0.9956240	0.0094709	-0.9953090	0.0094710	-0.9953090	1.90276×10^{-8}	7.75425×10^{-8}
	0.20	0.6103320	-0.8796460	0.6106390	-0.8811210	0.6106390	-0.8811210	2.69185×10^{-8}	1.81638×10^{-8}
0	0.40	0.5979280	-0.8788030	0.5986870	-0.8798700	0.5986880	-0.8798700	4.24645×10^{-7}	2.98579×10^{-7}
	0.60	0.5853820	-0.8782520	0.5866150	-0.8787530	0.5866180	-0.8787510	2.11781×10^{-6}	1.55116×10^{-6}
	0.20	0.9999710	-0.9999840	0.9999710	-0.9999860	0.9999710	-0.9999860	7.23998×10^{-12}	3.61844×10^{-12}
20	0.40	0.9999700	-0.9999830	0.9999700	-0.9999850	0.9999000	-0.9999850	1.17016×10^{-10}	5.84818×10^{-11}
	0.60	0.9999680	-0.9999820	0.9999680	-0.9999840	0.9999680	-0.9999840	5.98445×10^{-10}	2.99087×10^{-10}

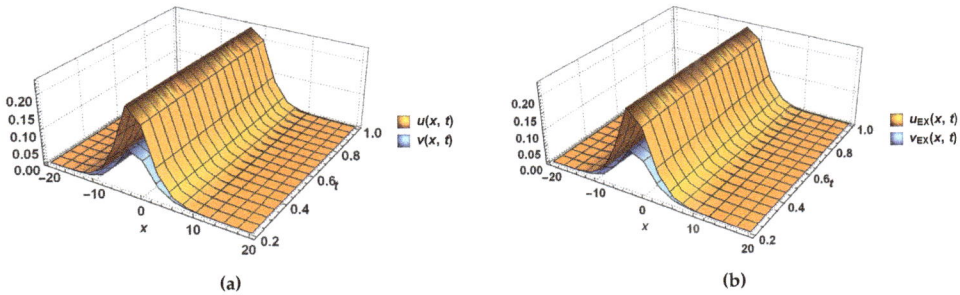

Figure 2. (**a**) The graph for the approximate solution of Example 1 for $\alpha = \beta = 0.5, \gamma = 1$ and $q = 1$; (**b**) The graph for the exact solution of Example 1 for $\alpha = \beta = 0.5$ and $\gamma = 1$.

5. Discussion and Conclusions

In this paper, a nonlinear time-fractional partial differential system via NAT was studied. Moreover, the convergence and error analysis was also shown. From the computational point of view, the solutions obtained by our technique were in excellent agreement with those obtained via previous works and also conformed with the exact solution to confirm the effectiveness and accuracy of this technique. Moreover, approximate traveling wave solutions for some systems of nonlinear wave equations were successfully obtained. We used Mathematica software to obtain the approximate and numerical results as well as drawing the graphs.

Acknowledgments: The work of the first author is supported by Savitribai Phule Pune University (formerly University of Pune), Pune 411007 (MS), India; The authors thanks the referees for their valuable suggestions and comments.

Author Contributions: Hayman Thabet and Subhash Kendre contributed substantially to this paper. Hayman Thabet wrote this paper; Subhash Kendre Supervised development of the paper and Dimplekumar Chalishajar helped to evaluate and edit the paper.

Conflicts of Interest: The authors declare no conflict of interest.

References

1. Baleanu, D. Special issue on nonlinear fractional differential equations and their applications in honour of Ravi P. Agarwal on his 65th birthday. *Nonlinear Dyn.* **2013**, *71*, 603.

2. Chalishajar, D.; George, R.; Nandakumaran, A.; Acharya, F. Trajectory controllability of nonlinear integro-differential system. *J. Franklin Inst.* **2010**, *347*, 1065–1075.

3. De Souza, M. On a class of nonhomogeneous fractional quasilinear equations in $\mathbb{R}^n$ with exponential growth. *Nonlinear Differ. Equ. Appl. NoDEA.* **2015**, *22*, 499–511.

4. Demir, A.; Erman, S.; Özgür, B.; Korkmaz, E. Analysis of fractional partial differential equations by Taylor series expansion. *Bound. Value Probl.* **2013**, *2013*, 68.

5. Huang, Q.; Zhdanov, R. Symmetries and exact solutions of the time fractional Harry-Dym equation with Riemann–Liouville derivative. *Phys. A Stat. Mech. Appl.* **2014**, *409*, 110–118.

6. Sun, H.; Chen, W.; Chen, Y. Variable-order fractional differential operators in anomalous diffusion modeling. *Phys. A Stat. Mech. Appl.* **2009**, *388*, 4586–4592.

7. Sun, H.; Chen, W.; Li, C.; Chen, Y. Fractional differential models for anomalous diffusion. *Phys. A Stat. Mech. Appl.* **2010**, *389*, 2719–2724.

8. Sun, H.G.; Chen, W.; Sheng, H.; Chen, Y.Q. On mean square displacement behaviors of anomalous diffusions with variable and random orders. *Phys. Lett. A* **2010**, *374*, 906–910.

9. Zayed, E.M.E.; Amer, Y.A.; Shohib, R.M.A. The fractional complex transformation for nonlinear fractional partial differential equations in the mathematical physics. *J. Assoc. Arab Univ. Basic Appl. Sci.* **2016**, *19*, 59–69.

10. Momani, S.; Odibat, Z. Homotopy perturbation method for nonlinear partial differential equations of fractional order. *Phys. Lett. A* **2007**, *365*, 345–350.

11. El-Sayed, A.; Elsaid, a.; El-Kalla, I.; Hammad, D. A homotopy perturbation technique for solving partial differential equations of fractional order in finite domains. *Appl. Math. Comput.* **2012**, *218*, 8329–8340.

12. Odibat, Z.; Momani, S. Numerical methods for nonlinear partial differential equations of fractional order. *Appl. Math. Model.* **2008**, *32*, 28–39.

13. Ma, W.X.; Lee, J.H. A transformed rational function method and exact solutions to the 3 + 1 dimensional Jimbo-Miwa equation. *Chaos Solitons Fractals* **2009**, *42*, 1356–1363.

14. Ma, W.X.; Zhu, Z. Solving the (3 + 1)-dimensional generalized KP and BKP equations by the multiple exp-function algorithm. *Appl. Math. Comput.* **2012**, *218*, 11871–11879.

15. Magin, R.; Ortigueira, M.D.; Podlubny, I.; Trujillo, J. On the fractional signals and systems. *Signal Proc.* **2011**, *91*, 350–371.

16. Mamchuev, M.O. Cauchy problem in nonlocal statement for a system of fractional partial differential equations. *Differ. Equ.* **2012**, *48*, 354–361.

17. Mamchuev, M.O. Mixed problem for a system of fractional partial differential equations. *Differ. Equ.* **2016**, *52*, 133–138.

18. Longuet-Higgins, M.; Stewart, R. Radiation stresses in water waves; a physical discussion, with applications. *Deep Sea Res. Oceanogr. Abstr.* **1964**, *11*, 529–562.

19. Slunyaev, A.; Didenkulova, I.; Pelinovsky, E. Rogue waters. *Contemp. Phys.* **2011**, *52*, 571–590.

20. Bai, C. New explicit and exact travelling wave solutions for a system of dispersive long wave equations. *Rep. Math. Phys.* **2004**, *53*, 291–299.

21. Benney, D.J.; Luke, J.C. On the Interactions of Permanent Waves of Finite Amplitude. *J. Math. Phys.* **1964**, *43*, 309–313.

22. Wang, M. Solitary wave solutions for variant Boussinesq equations. *Phys. Lett. A* **1995**, *199*, 169–172.
23. Lu, B. N-soliton solutions of a system of coupled KdV equations. *Phys. Lett. A* **1994**, *189*, 25–26.
24. Wang, M.; Zhou, Y.; Li, Z. Application of a homogeneous balance method to exact solutions of nonlinear equations in mathematical physics. *Phys. Lett. A.* **1996**, *216*, 67–75.
25. Lou, S.Y. Painlevé test for the integrable dispersive long wave equations in two space dimensions. *Phys. Lett. A* **1993**, *176*, 96–100.
26. Paquin, G.; Winternitz, P. Group theoretical analysis of dispersive long wave equations in two space dimensions. *Phys. D Nonlinear Phenom.* **1990**, *46*, 122–138.
27. Jafari, H.; Seifi, S. Solving a system of nonlinear fractional partial differential equations using homotopy analysis method. *Commun. Nonlinear Sci. Numer. Simul.* **2009**, *14*, 1962–1969.
28. Jafari, H.; Nazari, M.; Baleanu, D.; Khalique, C.M. A new approach for solving a system of fractional partial differential equations. *Comput. Math. Appl.* **2013**, *66*, 838–843.
29. Zhang, R.; Gong, J. Synchronization of the fractional-order chaotic system via adaptive observer. *Syst. Sci. Control Eng.* **2014**, *2*, 751–754.
30. Diethelm, K. *The Analysis of Fractional Differential Equations: An Application-Oriented Exposition Using Differential Operators of Caputo Type*; Springer: Berlin, Germany, 2010.
31. Ghorbani, A. Toward a new analytical method for solving nonlinear fractional differential equations. *Comput. Methods Appl. Mech. Eng.* **2008**, *197*, 4173–4179.
32. Guo, B.; Pu, X.; Huang, F. *Fractional Partial Differential Equations and Their Numerical Solutions*; World Scientific Publishing Co Pte Ltd: Singapore, 2015.
33. Kilbas, A.A.A.; Srivastava, H.M.; Trujillo, J.J. *Theory and Applications of Fractional Differential Equations*; Elsevier Science Limited: Amsterdam, Netherlands, 2006; Volume 204.
34. Miller, K.S.; Ross, B. *An Introduction to the Fractional Calculus and Fractional Differential Equations*; Wiley-Interscience: New York, NY, USA, 1993.
35. Podlubny, I. *Fractional Differential Equations: An Introduction to Fractional Derivatives, Fractional Differential Equations, to Methods of Their Solution and Some of Their Applications*; Academic Press: Salt Lake City, UT, USA, 1998; p. 198.

mathematics

MDPI

Correction

Correction: Thabet, H.; Kendre, S.; Chalishajar, D. New Analytical Technique for Solving a System of Nonlinear Fractional Partial Differential Equations *Mathematics* 2017, 5, 47

Hayman Thabet [1], Subhash Kendre [2] and Dimplekumar Chalishajar [2,*]

[1] Department of Mathematics, Savitribai Phule Pune University, Pune 411007, India;
    haymanthabet@gmail.com
[2] Department of Applied Mathematics, Virginia Military Institute, Lexington, VA 24450, USA;
    dipu17370@gmail.com
*   Correspondence: sdkendre@yahoo.com

Received: 13 February 2018; Accepted: 14 February 2018; Published: 14 February 2018

We have found some errors in the caption of Figure 1 and Figure 2 in our paper [1], and thus would like to make the following corrections:

On page 10, the caption of Figure 1 should be changed from:

Figure 1. (**a**) The graph for the approximate solution of Example 1 for $\alpha = \beta = 0.5$, $\gamma = 1$ and $q = 1$; (**b**) The graph for the exact solution of Example 1 for $\alpha = \beta = 0.5$ and $\gamma = 1$.

To the following correct version:

Figure 1. (**a**) The graph for the approximate solution of Example 2 for $\alpha = \beta = 0.5$ and $q_1 = q_2 = 1$; (**b**) The graph for the exact solution of Example 2 for $\alpha = \beta = 0.5$.

Furthermore, on page 13, the caption of Figure 2 should be changed from:

Figure 2. (**a**) The graph for the approximate solution of Example 2 for $\alpha = \beta = 0.5$ and $q_1 = q_2 = 1$; (**b**) The graph for the exact solution of Example 2 for $\alpha = \beta = 0.5$.

To the following correct version:

Figure 2. (**a**) The graph for the approximate solution of Example 1 for $\alpha = \beta = 0.5$, $\gamma = 1$ and $q = 1$; (**b**) The graph for the exact solution of Example 1 for $\alpha = \beta = 0.5$ and $\gamma = 1$.

The authors apologize for any inconvenience caused to the readers. The change does not affect the scientific results. The manuscript will be updated and the original will remain online on the article webpage.

References

1.   Thabet, H.; Kendre, S.; Chalishajar, D. New Analytical Technique for Solving a System of Nonlinear Fractional Partial Differential Equations. *Mathematics* **2017**, *5*, 47. [CrossRef]

mathematics

MDPI

Article

Stability of a Monomial Functional Equation on a Restricted Domain

Yang-Hi Lee

Department of Mathematics Education, Gongju National University of Education, Gongju 32553, Korea;
yanghi2@hanmail.net

Academic Editor: Hari Mohan Srivastava
Received: 24 August 2017; Accepted: 8 October 2017; Published: 18 October 2017

Abstract: In this paper, we prove the stability of the following functional equation $\sum_{i=0}^{n} {}_nC_i(-1)^{n-i}f(ix+y) - n!f(x) = 0$ on a restricted domain by employing the direct method in the sense of Hyers.

Keywords: stability; monomial functional equation; direct method

MSC: 39B82; 39B52

1. Introduction

Let V and W be real vector spaces, X a real normed space, Y a real Banach space, $n \in \mathbb{N}$ (the set of natural numbers), and $f : V \to W$ a given mapping. Consider the functional equation

$$\sum_{i=0}^{n} {}_nC_i(-1)^{n-i}f(ix+y) - n!f(x) = 0 \qquad (1)$$

for all $x, y \in V$, where ${}_nC_i := \frac{n!}{i!(n-i)!}$. The functional Equation (1) is called an n-monomial functional equation and every solution of the functional Equation (1) is said to be a monomial mapping of degree n. The function $f : \mathbb{R} \to \mathbb{R}$ given by $f(x) := ax^n$ is a particular solution of the functional Equation (1). In particular, the functional Equation (1) is called an additive (quadratic, cubic, quartic, and quintic, respectively) functional equation for the case $n = 1$ ($n = 2$, $n = 3$, $n = 4$, and $n = 5$, respectively) and every solution of the functional Equation (1) is said to be an additive (quadratic, cubic, quartic, and quintic, respectively) mapping for the case $n = 1$ ($n = 2$, $n = 3$, $n = 4$, and $n = 5$, respectively).

A mapping $A : V \to W$ is said to be additive if $A(x + y) = A(x) + A(y)$ for all $x, y \in V$. It is easy to see that $A(rx) = rA(x)$ for all $x \in V$ and all $r \in \mathbb{Q}$ (the set of rational numbers). A mapping $A_n : V^n \to W$ is called n-additive if it is additive in each of its variables. A mapping A_n is called symmetric if $A_n(x_1, x_2, \ldots, x_n) = A_n(x_{\pi(1)}, x_{\pi(2)}, \ldots, x_{\pi(n)})$ for every permutation $\pi : \{1, 2, \ldots, n\} \to \{1, 2, \ldots, n\}$. If $A_n(x_1, x_2, \ldots, x_n)$ is an n-additive symmetric mapping, then $A^n(x)$ will denote the diagonal $A_n(x, x, \ldots, x)$ for $x \in V$ and note that $A^n(rx) = r^n A^n(x)$ whenever $x \in V$ and $r \in \mathbb{Q}$. Such a mapping $A^n(x)$ will be called a monomial mapping of degree n (assuming $A^n \not\equiv 0$). Furthermore, the resulting mapping after substitution $x_1 = x_2 = \ldots = x_l = x$ and $x_{l+1} = x_{l+2} = \ldots = x_n = y$ in $A_n(x_1, x_2, \ldots, x_n)$ will be denoted by $A^{l,n-l}(x, y)$. A mapping $p : V \to W$ is called a generalized polynomial (GP) mapping of degree $n \in N$ provided that there exist $A^0(x) = A^0 \in W$ and i-additive symmetric mappings $A^i : V^i \to W$ (for $1 \leq i \leq n$) such that $p(x) = \sum_{i=0}^{n} A^i(x)$, for all $x \in V$ and $A^n \not\equiv 0$. For $f : V \to W$, let Δ_h be the difference operator defined as follows:

$$\Delta_h f(x) = f(x + h) - f(x)$$

for $h \in V$. Furthermore, let $\Delta_h^0 f(x) = f(x)$, $\Delta_h^1 f(x) = \Delta_h f(x)$ and $\Delta_h \circ \Delta_h^n f(x) = \Delta_h^{n+1} f(x)$ for all $n \in N$ and all $h \in V$. For any given $n \in N$, the functional equation $\Delta_h^{n+1} f(x) = 0$ for all $x, h \in V$ is well studied. In explicit form we can have

$$\Delta_h^n f(x) = \sum_{i=0}^{n} {}_nC_i(-1)^{n-i} f(x + ih).$$

The following theorem was proved by Mazur and Orlicz [1,2] and in greater generality by Djoković (see [3]).

Theorem 1. *Let V and W be real vector spaces, $n \in \mathbb{N}$ and $f : V \to W$, then the following are equivalent:*

(1) $\Delta_h^{n+1} f(x) = 0$ *for all $x, h \in V$.*
(2) $\Delta_{x_1} \circ \Delta_{x_2} \circ \ldots \circ \Delta_{x_{n+1}} f(x_0) = 0$ *for all $x_0, x_1, x_2, \ldots, x_{n+1} \in V$.*
(3) $f(x) = A^n(x) + A^{n-1}(x) + \cdots + A^2(x) + A^1(x) + A^0(x)$ *for all $x \in V$, where $A^0(x) = A^0$ is an arbitrary element of W and $A^i(x)(i = 1, 2, \ldots, n)$ is the diagonal of an i-additive symmetric mapping $A^i : V^i \to W$.*

In 2007, L. Cădariu and V. Radu [4] proved a stability of the monomial functional Equation (1) (see also [5–7]), in particular, the following result is given by the author in [6].

Theorem 2. *Let p be a non-negative real number with $p \neq n$, let $\theta > 0$, and let $f : X \to Y$ be a mapping such that*

$$\left\| \sum_{i=0}^{n} {}_nC_i(-1)^{n-i} f(ix + y) - n! f(x) \right\| \leq \theta(\|x\|^p + \|y\|^p) \tag{2}$$

for all $x, y \in X$. Then there exist a positive real number K and a unique monomial function of degree n $F : X \to Y$ such that

$$\|f(x) - F(x)\| \leq K\|x\|^p \tag{3}$$

holds for all $x \in X$. The mapping $F : X \to Y$ is given by

$$F(x) := \lim_{s \to \infty} \frac{f(2^s x)}{2^{ns}}$$

for all $x \in X$.

The concept of stability for the functional Equation (1) arises when we replace the functional Equation (1) by an inequality (2), which is regarded as a perturbation of the equation. Thus, the stability question of functional Equation (1) is whether there is an exact solution of (1) near each solution of inequality (2). If the answer is affirmative with inequality (3), we would say that the Equation (1) is stable.

The direct method of Hyers means that, in Theorem 2, $F(x)$ satisfying inequality (3) is constructed by the limit of the sequence $\{\frac{f(2^s x)}{2^{ns}}\}_{s \in \mathbb{N}}$ as $s \to \infty$.

Historically, in 1940, Ulam [8] proposed the problem concerning the stability of group homomorphisms. In 1941, Hyers [9] gave an affirmative answer to this problem for additive mappings between Banach spaces, using the direct method. Subsequently, many mathematicians came to deal with this problem (cf. [10–17]).

In 1998, A. Gilányi dealt with the stability of monomial functional equation for the case $p = 0$ (see [18,19]) and he proved for the case when p is a real constant (see [20]). Thereafter, C.-K. Choi proved stability theorems for many kinds of restricted domains, but his theorems are mainly connected with the case of $p = 0$. If $p < 0$ in (2), then the inequality (2) cannot hold for all $x \in X$, so we have to restrict the domain by excluding 0 from X.

The main purpose of this paper is to generalize our previous result (Theorem 2) by replacing the real normed space X with a restricted domain S of a real vector space V and by replacing the control function $\theta(||x||^p + ||y||^p)$ with a more general function $\varphi : S^2 \to [0, \infty)$.

2. Stability of the Functional Equation (1) on a Restricted Domain

In this section, for a given mapping $f : V \to W$, we use the following abbreviation

$$D_n f(x, y) := \sum_{i=0}^{n} {}_nC_i (-1)^{n-i} f(ix + y) - n! f(x)$$

for all $x, y \in V$.

Lemma 1. *The equalities*

$$
{}_nC_i = \sum_{\substack{j+k=2i \\ 0 \le j,k \le n}} {}_nC_j \cdot {}_nC_k (-1)^{i+k} \quad (i = 0, \cdots, n) \tag{4}
$$

and

$$
{}_nC_i = \sum_{\substack{j+k=2i-1 \\ 0 \le j,k \le n}} {}_nC_j \cdot {}_nC_k (-1)^{n-k} \quad (i = 1, \cdots, n) \tag{5}
$$

hold.

Proof. From the equalities

$$\sum_{i=0}^{n} {}_nC_i (-1)^{n-i} x^{2i} = (x^2 - 1)^n,$$

$$(x^2 - 1)^n = (x + 1)^n (x - 1)^n,$$

$$(x+1)^n (x-1)^n = \left(\sum_{j=0}^{n} {}_nC_j x^j \right) \left(\sum_{k=0}^{n} {}_nC_k (-1)^{n-k} x^k \right) = \sum_{j=0}^{n} \sum_{k=0}^{n} {}_nC_j \cdot {}_nC_k (-1)^{n-k} x^{j+k},$$

we get the equality

$$\sum_{i=0}^{n} {}_nC_i (-1)^{n-i} x^{2i} = \sum_{j=0}^{n} \sum_{k=0}^{n} {}_nC_j \cdot {}_nC_k (-1)^{n-k} x^{j+k} \tag{6}$$

for all $x \in \mathbb{R}$. Since the coefficient of the term x^{2i} of the left-hand side in (6) is ${}_nC_i (-1)^{n-i}$ and the coefficient of the term x^{2i} of the right-hand side in (6) is $\sum_{\substack{j+k=2i \\ 0 \le j,k \le n}} {}_nC_j \cdot {}_nC_k (-1)^{n-k}$, we get the Equality (4).

We easily know that the coefficient of the term x^{2i-1} of the left-hand side in (6) is 0 and the coefficient of the term x^{2i-1} of the right-hand side in (6) is $\sum_{\substack{j+k=2i-1 \\ 0 \le j,k \le n}} {}_nC_j \cdot {}_nC_k (-1)^{n-k}$. So we get the Equality (5). $\square$

We rewrite a refinement of the result given in [6].

Lemma 2. (Lemma 1 in [6]) *The equality*

$$\sum_{j=0}^{n} {}_nC_j D_n f(x, jx + y) - D_n f(2x, y) = n!(f(2x) - 2^n f(x)) \tag{7}$$

holds for all $x, y \in V$. *In particular, if* $D_n f(x, y) = 0$ *for all* $x, y \in V$, *then*

$$f(2x) = 2^n f(x) \tag{8}$$

Proof. By (4), (5), and the equality $\sum_{j=0}^{n} {}_nC_j = 2^n$, we get the equalities

$$\sum_{j=0}^{n} {}_nC_j D_n f(x, jx + y) - D_n f(2x, y)$$

$$= \sum_{j=0}^{n} {}_nC_j \sum_{k=0}^{n} (-1)^{n-k} {}_nC_k f((j+k)x + y) - \sum_{j=0}^{n} {}_nC_j n! f(x)$$

$$- \sum_{i=0}^{n} {}_nC_i (-1)^{n-i} f(2ix + y) + n! f(2x)$$

$$= \sum_{i=0}^{n} \sum_{\substack{j+k=2i \\ 0 \le j,k \le n}} {}_nC_j (-1)^{n-k} {}_nC_k f(2ix + y)$$

$$+ \sum_{i=1}^{n} \sum_{\substack{j+k=2i-1 \\ 0 \le j,k \le n}} {}_nC_j (-1)^{n-k} {}_nC_k f((2i-1)x + y)$$

$$- \sum_{j=0}^{n} {}_nC_j n! f(x) - \sum_{i=0}^{n} {}_nC_i (-1)^{n-i} f(2ix + y) + n! f(2x)$$

$$= - \sum_{j=0}^{n} {}_nC_j n! f(x) + n! f(2x)$$

for all $x, y \in V$. $\qquad \square$

Lemma 3. *If* f *satisfies the functional equation* $D_n f(x, y) = 0$ *for all* $x, y \in V \backslash \{0\}$ *with* $f(0) = 0$, *then* f *satisfies the functional equation* $D_n f(x, y) = 0$ *for all* $x, y \in V$.

Proof. Since $D_n f(0, 0) = 0$, $D_n f(0, y) = 0$ for all $y \in V \backslash \{0\}$, and

$$D_n f(x, 0) = (-1)^n D_n f(-x, nx) - (-1)^n D_n f(-x, (n+1)x) + D_n f(x, x)$$

for all $x \in V \backslash \{0\}$, we conclude that f satisfies the functional equation $D_n f(x, y) = 0$ for all $x, y \in V$. $\square$

We rewrite a refinement of the result given in [7].

Theorem 3. (*Corollary 4 in* [7]) *A mapping* $f : V \to W$ *is a solution of the functional Equation* (1) *if and only if* f *is of the form* $f(x) = A^n(x)$ *for all* $x \in V$, *where* A^n *is the diagonal of the n-additive symmetric mapping* $A_n : V^n \to W$.

Proof. Assume that f satisfies the functional Equation (1). We get the equation $\Delta_x^{n+1} f(y) = D_n f(x, x + y) - D_n f(x, y) = 0$ for all $x, y \in V$. By Theorem 1, f is a generalized polynomial mapping of degree at most n, that is, f is of the form $f(x) = A^n(x) + A^{n-1}(x) + \cdots + A^2(x) + A^1(x) + A^0(x)$ for all $x \in V$, where $A^0(x) = A^0$ is an arbitrary element of W and $A^i(x) (i = 1, 2, \ldots, n)$ is the diagonal of an i-additive symmetric mapping $A_i : V^i \to W$. On the other hand, $f(2x) = 2^n f(x)$ holds for all $x \in V$ by Lemma 2, and so $f(x) = A^n(x)$.

Conversely, assume that $f(x) = A^n(x)$ for all $x \in V$, where $A^n(x)$ is the diagonal of the n-additive symmetric mapping $A_n : V^n \to W$. From $A^n(x + y) = A^n(x) + \sum_{i=1}^{n-1} {}_nC_i A^{n-i,i}(x, y)$,

$A^n(rx) = r^n A^n(x)$, $A^{n-i,i}(x, ry) = r^i A^{n-i,i}(x, y)$ $(x, y \in V, r \in \mathbb{Q})$, we see that f satisfies (1), which completes the proof of this theorem. $\qquad\square$

Theorem 4. *Let S be a subset of a real vector space V and Y a real Banach space. Suppose that for each $x \in V \backslash \{0\}$ there exists a real number $r_x > 0$ such that $rx \in S$ for all $r \geq r_x$. Let $\varphi : S^2 \to [0, \infty)$ be a function such that*

$$\sum_{k=0}^{\infty} 2^{-nk} \varphi(2^k x, 2^k y) < \infty \tag{9}$$

for all $x, y \in S$. If the mapping $f : S \to Y$ satisfies the inequality

$$\|D_n f(x, y)\| \leq \varphi(x, y) \tag{10}$$

for all $x, y \in S$, then there exists a unique monomial mapping of degree n $F : V \to Y$ such that

$$\|f(x) - F(x)\| \leq \sum_{k=0}^{\infty} \frac{\Phi(2^k x)}{n! \cdot 2^{(k+1)n}} \tag{11}$$

for all $x \in S$, where

$$\Phi(x) := \sum_{j=0}^{n} {}_nC_j \varphi(x, jx + x) + \varphi(2x, x).$$

In particular, F is represented by

$$F(x) = \lim_{m \to \infty} \frac{f(2^m x)}{2^{mn}}$$

for all $x \in V$.

Proof. Let $x \in V \backslash \{0\}$ and m be an integer such that $2^m \geq r_x$. It follows from (7) in Lemma 2 and (10) that

$$
\begin{aligned}
n! \left\| f(2^{m+1}x) - 2^n f(2^m x) \right\| &= \left\| \sum_{j=0}^{n} {}_nC_j D_n f(2^m x, 2^m(jx + x))) - D_n f(2^{m+1}x, 2^m x) \right\| \\
&\leq \sum_{j=0}^{n} \left\| {}_nC_j D_n f(2^m x, 2^m(j+1)x) \right\| + \| D_n f(2^{m+1}x, 2^m x) \| \\
&\leq \sum_{j=0}^{n} {}_nC_j \varphi(2^m x, 2^m(j+1)x) + \varphi(2^{m+1}x, 2^m x) \\
&= \Phi(2^m x).
\end{aligned}
$$

From the above inequality, we get the following inequalities

$$\left\| \frac{f(2^m x)}{2^{mn}} - \frac{f(2^{m+1}x)}{2^{(m+1)n}} \right\| \leq \frac{\Phi(2^m x)}{n! \cdot 2^{n(m+1)}}$$

and

$$\left\| \frac{f(2^m x)}{2^{nm}} - \frac{f\left(2^{m+m'} x\right)}{2^{n(m+m')}} \right\| \leq \sum_{k=m}^{m+m'-1} \frac{\Phi(2^k x)}{n! \cdot 2^{n(k+1)}} \tag{12}$$

for all $m' \in \mathbb{N}$. So the sequence $\{\frac{f(2^m x)}{2^{mn}}\}_{m \in \mathbb{N}}$ is a Cauchy sequence for all $x \in V \backslash \{0\}$. Since $\lim_{m \to \infty} \frac{f(2^m 0)}{2^{mn}} = 0$ and Y is a real Banach space, we can define a mapping $F : V \to Y$ by

$$F(x) = \lim_{m \to \infty} \frac{f(2^m x)}{2^{nm}}$$

for all $x \in V$. By putting $m = 0$ and letting $m' \to \infty$ in the inequality (12), we obtain the inequality (11) if $x \in S$.

From the inequality (10), we get

$$\left\| \frac{D_n f(2^m x, 2^m y)}{2^{nm}} \right\| \le \frac{\varphi(2^m x, 2^m y)}{2^{nm}}.$$

for all $x, y \in V \backslash \{0\}$, where $2^m \ge r_x, r_y$. Since the right-hand side in the above equality tends to zero as $m \to \infty$, we obtain that F satisfies the inequality (1) for all $x, y \in V \backslash \{0\}$. By Lemma 3 and $F(0) = 0$, F satisfies the Equality (1) for all $x, y \in V$. To prove the uniqueness of F, assume that F' is another monomial mapping of degree n satisfying the inequality (11) for all $x \in S$. The equality $F'(x) = \frac{F'(2^m x)}{2^{nm}}$ follows from the Equality (8) in Lemma 2 for all $x \in V \backslash \{0\}$ and $m \in \mathbb{N}$. Thus we can obtain the inequalities

$$\left\| F'(x) - \frac{f(2^m x)}{2^{nm}} \right\| = \left\| \frac{F'(2^m x)}{2^{nm}} - \frac{f(2^m x)}{2^{nm}} \right\| \le \sum_{k=0}^{\infty} \frac{\Phi(2^{k+m} x)}{n! \cdot 2^{n(k+m+1)}} = \sum_{k=m}^{\infty} \frac{\Phi(2^k x)}{n! \cdot 2^{n(k+1)}},$$

for all $x \in V \backslash \{0\}$, where $2^m \ge r_x$. Since $\sum_{k=m}^{\infty} \frac{\Phi(2^k x)}{n! \cdot 2^{n(k+1)}} \to 0$ as $m \to \infty$ and $F'(0) = 0$, $F'(x) = \lim_{m \to \infty} 2^{-nm} f(2^m x)$ for all $x \in V$, i.e., $F(x) = F'(x)$ for all $x \in V$. This completes the proof of the theorem. $\square$

We can give a generalization of Theorems 2 and 5 in [6] as the following corollary.

Corollary 1. *Let p and r be real numbers with $p < n$ and $r > 0$, let X be a normed space, $\varepsilon > 0$, and $f : X \to Y$ be a mapping such that*

$$\|D_n f(x, y)\| \le \varepsilon(\|x\|^p + \|y\|^p) \tag{13}$$

for all $x, y \in X$ with $\|x\|, \|y\| > r$. Then there exists a unique monomial mapping of degree n $F : X \to Y$ satisfying

$$\|f(x) - F(x)\| \le \begin{cases} \frac{\varepsilon(2^p + 2^n + 1 + \sum_{j=0}^{n} {}_n C_j (j+1)^p)}{n!(2^n - 2^p)} \|x\|^p & (\text{for } \|x\| > r), \\ \frac{((k+1)^p + k^p)\varepsilon \|x\|^p}{n} + (k^p + \sum_{i=2}^{n} {}_n C_i (ik + i - k)^p + n!(k+1)^p) \\ \times \frac{\varepsilon(2^p + 2^n + 1 + \sum_{j=0}^{n} {}_n C_j (j+1)^p)}{n! n (2^n - 2^p)} \|x\|^p & (\text{for } \|kx\| > r). \end{cases}$$

In particular, if $p < 0$, then f is a monomial mapping of degree n itself.

Proof. If we set $\varphi(x, y) := \varepsilon(\|x\|^p + \|y\|^p)$ and $S = \{x \in X | \|x\| > r\}$, then there exists a unique monomial mapping of degree n $F : X \to Y$ satisfying

$$\|f(x) - F(x)\| \le \frac{\varepsilon(2^p + 2^n + 1 + \sum_{j=0}^{n} {}_n C_j (j+1)^p)}{n!(2^n - 2^p)} \|x\|^p \tag{14}$$

for all $x \in X$ with $\|x\| > r$ by Theorem 4. Notice that if F is a monomial mapping of degree n, then $D_n F((k+1)x, -kx) = 0$ for all $x \in X$ and $k \in \mathbb{R}$. Hence the equality

$$
\begin{aligned}
(-1)^n \cdot n(f(x) - F(x)) = {} & D_n f((k+1)x, -kx) + (-1)^n (F - f)(-kx) \\
& + \sum_{i=2}^{n} {}_n C_i (-1)^{n-i} (F - f)(i(k+1)x - kx) \\
& - n! (F - f)((k+1)x)
\end{aligned}
$$

holds for all $x \in X$ and $k \in \mathbb{R}$. So if $F : X \to Y$ is the monomial mapping of degree n satisfying (14), then $F : X \to Y$ satisfies the inequality with a real number k

$$
\begin{aligned}
\|f(x) - F(x)\| \leq {} & \frac{((k+1)^p + k^p)\varepsilon \|x\|^p}{n} + \left(k^p + \sum_{i=2}^{n} {}_n C_i (ik + i - k)^p + n!(k+1)^p \right) \\
& \times \frac{\varepsilon(2^p + 2^n + 1 + \sum_{j=0}^{n} {}_n C_j (j+1)^p)}{n! n (2^n - 2^p)} \|x\|^p
\end{aligned} \tag{15}
$$

for all $\|kx\| > r$.

Moreover, if $p < 0$, then $\lim_{k \to \infty} (k^p + \sum_{i=2}^{n} C_i^n (ik + i - k)^p + n!(k+1)^p) = 0$ and $\lim_{k \to \infty} (k^p + (k+1)^p) = 0$. Hence we get

$$
f(x) = F(x)
$$

for all $x \in X \backslash \{0\}$ by (15). Since $\lim_{k \to \infty} k^p = 0$ and the inequality

$$
\begin{aligned}
\|f(0) - F(0)\| \leq {} & \frac{1}{n} \| D_n(f - F)(kx, -kx) + (-1)^n (F - f)(-kx) \\
& + \sum_{i=2}^{n} {}_n C_i (-1)^{n-i} (F - f)((i-1)kx) - n!(F - f)(kx) \| \\
\leq {} & \frac{1}{n} \left[2 + \frac{2^p + 2^n}{n!(2^n - 2^p)} \left(1 + \sum_{i=1}^{n-1} {}_n C_{i+1} i^p + n! \right) \right] k^p \varepsilon \|x\|^p
\end{aligned}
$$

holds for any fixed $x \in X \backslash \{0\}$ with $\|x\| > r$ and all natural numbers k, we get

$$
f(0) = F(0).
$$

$\square$

Theorem 5. *Let S be a subset of a real vector space V and Y a real Banach space. Suppose that for each $x \in V$ there exists a real number $r_x > 0$ such that $rx \in S$ for all $r \leq r_x$. Let $\varphi : V^2 \to [0, \infty)$ be a function such that*

$$
\sum_{k=0}^{\infty} 2^{nk} \varphi(2^{-k} x, 2^{-k} y) < \infty \tag{16}
$$

for all $x, y \in S$. Suppose that a mapping $f : V \to Y$ satisfies the inequality (10) for all $x, y \in S$, where $ix + y \in S$ for all $i = 0, 1, \ldots, n$. Then there exists a unique monomial mapping of degree n $F : V \to Y$ such that

$$
\|f(x) - F(x)\| \leq \sum_{k=0}^{\infty} \frac{2^{nk}}{n!} \Phi \left(\frac{x}{2^{k+1}} \right) \tag{17}
$$

for all x with $nx \in S$, where $\Phi(x)$ is defined as in Theorem 4. In particular, F is represented by

$$
F(x) = \lim_{m \to \infty} 2^{nm} f(2^{-m} x)
$$

for all $x \in V$.

Proof. Let $x \in V$ and m be an integer such that $2^{-m}(n+1) \leq r_x$. It follows from (7) in Lemma 2 and (10) that

$$n! \left\| f\left(\frac{x}{2^m}\right) - 2^n f\left(\frac{x}{2^{m+1}}\right) \right\| = \left\| \sum_{j=0}^{n} {}_nC_j D_n f\left(\frac{x}{2^{m+1}}, \frac{(j+1)x}{2^{m+1}}\right) - D_n f\left(\frac{x}{2^m}, \frac{x}{2^{m+1}}\right) \right\|$$

$$\leq \sum_{j=0}^{n} {}_nC_j \left\| D_n f\left(\frac{x}{2^{m+1}}, \frac{(j+1)x}{2^{m+1}}\right) \right\| + \left\| D_n f\left(\frac{x}{2^m}, \frac{x}{2^{m+1}}\right) \right\|$$

$$\leq \sum_{j=0}^{n} {}_nC_j \varphi\left(\frac{x}{2^{m+1}}, \frac{(j+1)x}{2^{m+1}}\right) + \varphi\left(\frac{x}{2^m}, \frac{x}{2^{m+1}}\right)$$

$$= \Phi\left(\frac{x}{2^{m+1}}\right)$$

for all $x \in V$. From the above inequality, we get the inequality

$$\left\| 2^{nm} f\left(\frac{x}{2^m}\right) - 2^{n(m+m')} f\left(\frac{x}{2^{m+m'}}\right) \right\| \leq \sum_{k=m}^{m+m'-1} \frac{2^{nk}}{n!} \Phi\left(\frac{x}{2^{k+1}}\right) \qquad (18)$$

for all $x \in V$ and $m' \in \mathbb{N}$. So the sequence $\{2^{nm} f\left(\frac{x}{2^m}\right)\}_{m \in \mathbb{N}}$ is a Cauchy sequence by the inequality (16). From the completeness of Y, we can define a mapping $F : V \to Y$ by

$$F(x) = \lim_{m \to \infty} 2^{nm} f\left(\frac{x}{2^m}\right)$$

for all $x \in V$. Moreover, by putting $m = 0$ and letting $m' \to \infty$ in (18), we get the inequality (17) for all $x \in S$ with $(n+1)x \in S$. From the inequality (10), if m is a positive integer such that $\frac{ix+y}{2^m} \in S$ for all $i = 0, 1, \ldots, n$, then we get

$$\left\| 2^{nm} D_n f\left(\frac{x}{2^m}, \frac{y}{2^m}\right) \right\| \leq 2^{nm} \varphi\left(\frac{x}{2^m}, \frac{y}{2^m}\right),$$

for all $x, y \in V$. Since the right-hand side in this inequality tends to zero as $m \to \infty$, we obtain that F is a monomial mapping of degree n. To prove the uniqueness of F assume that F' is another monomial mapping of degree n satisfying the inequality (17) for all $x \in S$ with $(n+1)x \in S$. So the equality $F'(x) = 2^{nm} F'\left(\frac{x}{2^m}\right)$ holds for all $x \in V$ by (8) in Lemma 2. Thus, we can infer that

$$\left\| F'(x) - 2^{nm} f\left(\frac{x}{2^m}\right) \right\| = \left\| 2^{nm} F'\left(\frac{x}{2^m}\right) - 2^{nm} f\left(\frac{x}{2^m}\right) \right\|$$

$$\leq \sum_{k=0}^{\infty} 2^{n(m+k)} \Phi\left(\frac{x}{2^{m+k+1}}\right)$$

$$\leq \sum_{k=m}^{\infty} 2^{nk} \Phi\left(\frac{x}{2^{k+1}}\right)$$

for all positive integers m, where $\frac{(n+1)x}{2^m} \leq r_x$. Since $\sum_{k=m}^{\infty} 2^{nk} \Phi\left(\frac{x}{2^{k+1}}\right) \to 0$ as $m \to \infty$, we know that $F'(x) = \lim_{m \to \infty} 2^{nm} f\left(\frac{x}{2^m}\right)$ for all $x \in V$. This completes the proof of the theorem. $\square$

We can give a generalization of Theorem 3 in [6] as the following corollary.

Mathematics **2017**, 5, 53

Corollary 2. *Let p and r be real numbers with $p > n$ and $r > 0$, and X a normed space. Let $f : X \to Y$ be a mapping satisfying the inequality* (13) *for all x, y with $\|x\|, \|y\| < r$. Then there exists a unique monomial mapping of degree n $F : X \to Y$ satisfying*

$$\|f(x) - F(x)\| \leq \frac{\varepsilon(2^p + 2^n + 1 + \sum_{j=0}^{n} {}_nC_j (j+1)^p)}{n!(2^n - 2^p)} \|x\|^p$$

for all $x \in X$ with $\|x\| < \frac{r}{n+1}$.

The following example shows that the assumption $p \neq n$ cannot be omitted in Corollarys 1 and 2. This example is an extension of the example of Gajda [21] for the monomial functional inequality (13) (see also [22]).

Example 1. *Let $\psi : \mathbb{R} \to \mathbb{R}$ be defined by*

$$\psi(x) = \begin{cases} x^n, & \text{for } |x| < 1, \\ 1, & \text{for } |x| \geq 1. \end{cases} \tag{19}$$

Consider that the function $f : \mathbb{R} \to \mathbb{R}$ is defined by

$$f(x) = \sum_{m=0}^{\infty} (n+1)^{-nm} \psi((n+1)^m x) \tag{20}$$

for all $x \in \mathbb{R}$. Then f satisfies the functional inequality

$$\left| \sum_{i=0}^{n} {}_nC_i (-1)^i f(ix + y) - n! f(x) \right| \leq 4 \cdot (n+1)!(n+1)^{2n}(|x|^n + |y|^n). \tag{21}$$

for all $x, y \in \mathbb{R}$, but there do not exist a monomial mapping of degree n $F : \mathbb{R} \to \mathbb{R}$ and a constant $d > 0$ such that $|f(x) - F(x)| \leq d|x|^n$ for all $x \in \mathbb{R}$.

Proof. It is clear that f is bounded by 2 on $\mathbb{R}$. If $|x|^n + |y|^n = 0$, then f satisfies (21). And if $|x|^n + |y|^n \geq \frac{1}{(n+1)^n}$, then

$$|D_n f(x, y)| \leq 2 \cdot 2 \cdot (n+1)! \leq 4 \cdot (n+1)!(n+1)^n(|x|^n + |y|^n),$$

which means that f satisfies (21). Now suppose that $0 < |x|^n + |y|^n < \frac{1}{(n+1)^n}$. Then there exists a nonnegative integer k such that

$$\frac{1}{(n+1)^{n(k+2)}} \leq |x|^n + |y|^n < \frac{1}{(n+1)^{n(k+1)}}. \tag{22}$$

Hence $(n+1)^{nk}|x|^n < \frac{1}{(n+1)^n}$, $(n+1)^{nk}|y|^n < \frac{1}{(n+1)^n}$, $|(n+1)^m(x+iy)| < 1$, and $|(n+1)^m y| < 1$ for all $m = 0, 1, \ldots, k-1$. Hence, for $m = 0, 1, \ldots, k-1$,

$$\sum_{i=0}^{n} {}_nC_i (-1)^i \psi((n+1)^m(ix+y)) - n! \psi((n+1)^m x) = 0. \tag{23}$$

From the definition of f, the inequality (22), and the inequality (23), we obtain that

$$
\begin{aligned}
|D_n f(x,y)| &= \left| \sum_{m=0}^{\infty} (n+1)^{-nm} \left(\sum_{i=0}^{n} {}_nC_i(-1)^i \psi((n+1)^m(ix+y)) - n!\psi((n+1)^m x) \right) \right| \\
&\leq \sum_{m=0}^{\infty} (n+1)^{-nm} \left| \sum_{i=0}^{n} {}_nC_i(-1)^i \psi((n+1)^m(ix+y)) - n!\psi((n+1)^m x) \right| \\
&\leq \sum_{m=k}^{\infty} (n+1)^{-nm} \left| \sum_{i=0}^{n} {}_nC_i(-1)^i \psi((n+1)^m(ix+y)) - n!\psi((n+1)^m x) \right| \\
&\leq \sum_{m=k}^{\infty} (n+1)^{-nm} 2 \cdot (n+1)! \leq 4 \cdot (n+1)^{-nk}(n+1)! \\
&\leq 4 \cdot (n+1)^{2n}(n+1)!(|x|^n + |y|^n).
\end{aligned}
\tag{24}
$$

Therefore, f satisfies (21) for all $x, y \in \mathbb{R}$. Now, we claim that the functional Equation (1) is not stable for $p = n$ in Corollarys 1 and 2. Suppose on the contrary that there exists a monomial mapping of degree n $F : R \to R$ and constant $d > 0$ such that $|f(x) - F(x)| \leq d|x|^n$ for all $x \in R$. Notice that $F(x) = x^n F(1)$ for all rational numbers x. So we obtain that

$$
|f(x)| \leq (d + |F(1)|)|x|^n
\tag{25}
$$

for all $x \in Q$. Let $k \in N$ with $k + 1 > d + |F(1)|$. If x is a rational number in $(0, (n+1)^{-k})$, then $(n+1)^m x \in (0,1)$ for all $m = 0, 1, \ldots, k$, and for this x we get

$$
\begin{aligned}
f(x) &= \sum_{m=0}^{\infty} (n+1)^{-nm} \psi((n+1)^m x) \geq \sum_{m=0}^{k} (n+1)^{-nm}((n+1)^m x)^n \\
&= (k+1)x^n > (d + |F(1)|)x^n
\end{aligned}
$$

which contradicts (25). □

3. Conclusions

The advantage of this paper is that we do not need to prove the stability of additive quadratic, cubic, and quartic functional equations separately. Instead we can apply our main theorem to prove the stability of those functional equations simultaneously.

Acknowledgments: The author would like to thank referees for their valuable suggestions and comments.

Conflicts of Interest: The author declares no conflict of interest.

References

1. Mazur, S.; Orlicz, W. Grundlegende eigenschaften der polynomischen operationen. Erste mitteilung. *Stud. Math.* **1934**, *5*, 50–68.
2. Mazur, S.; Orlicz, W. Grundlegende eigenschaften der polynomischen operationen. Zweite mitteilung. *Stud. Math.* **1934**, *5*, 179–189.
3. Djoković, D.Z. A representation theorem for $(X_1 - 1)(X_2 - 1) \cdots (X_n - 1)$ and its applications. *Ann. Polon. Math.* **1969** *22*, 189–198.
4. Cădariu, L.; Radu, V. The fixed points method for the stability of some functional equations. *Caepathian J. Math.* **2007**, *23*, 63–72.
5. Choi, C.-K. Stability of a Monomial Functional Equation on Restricted Domains of Lebesgue Measure Zero. *Results Math.* **2017**, 1–11, doi:10.1007/s00025-017-0742-0.
6. Lee, Y.-H. On the stability of the monomial functional equation. *Bull. Korean Math. Soc.* **2008**, *45*, 397–403.

7. Mckiernan, M.A. On vanishing *n*-th odered difference and Hamel bases, *Ann. Polon. Math.* **1967**, *19*, 331–336.
8. Ulam, S.M. *A Collection of Mathematical Problems*; Interscience Publishers: New York, NY, USA, 1960.
9. Hyers, D.H. On the stability of the linear functional equation. *Proc. Natl. Acad. Sci. USA* **1941**, *27*, 222–224.
10. Aoki, T. On the stability of the linear transformation in Banach spaces. *J. Math. Soc. Jpn.* **1950**, *2*, 64–66.
11. Cholewa, P.W. Remarks on the stability of functional equations. *Aequ. Math.* **1984**, *27*, 76–86.
12. Czerwik, S. *Functional Equations and Inequalities in Several Variables*; World Scientific: Hackensacks, NJ, USA, 2002.
13. Găvruta, P. A generalization of the Hyers–Ulam–Rassias stability of approximately additive mappings. *J. Math. Anal. Appl.* **1994**, *184*, 431–436.
14. Jung, S.-M. On the Hyers-Ulam stability of the functional equations that have the quadratic property. *J. Math. Anal. Appl.* **1998**, *222*, 126–137.
15. Lee, Y.H.; Jun, K.W. On the stability of approximately additive mappings. *Proc. Am. Math. Soc.* **2000**, *128*, 1361–1369.
16. Skof, F. Local properties and approximations of operators. *Rend. Sem. Mat. Fis. Milano* **1983**, *53*, 113–129.
17. Rassias, T.M. On the stability of the linear mapping in Banach spaces. *Proc. Am. Math. Soc.* **1978**, *72*, 297–300.
18. Gilányi, A. On Hyers-Ulam stability of monomial functional equations. *Abh. Math. Sem. Univ. Hamburg* **1998**, *68*, 321–328.
19. Gilányi, A. Hyers-Ulam stability of monomial functional equations on a general domain. *Proc Natl. Acad. Sci. USA* **1999**, *96*, 10588–10590.
20. Gilányi, A. On the stability of monomial functional equations. *Publ. Math. Debrecen* **2000**, *56*, 201–212.
21. Gajda, Z. On stability of additive mappings. *Int. J. Math. Math. Sci.* **1991**, *14*, 431–434.
22. Xu, T.Z.; Rassias, J.M.; Rassias, M.J.; Xu, W.X. A fixed point approach to the stability of quintic and sextic functional equations in quasi—Normed spaces. *J. Inequal. Appl.* **2010**, *2010*, 423231.

Article

An Investigation of Radial Basis Function-Finite Difference (RBF-FD) Method for Numerical Solution of Elliptic Partial Differential Equations

Suranon Yensiri [1] and Ruth J. Skulkhu [1,2,*]

[1] Department of Mathematics, Faculty of Science, Mahidol University, Bangkok 10400, Thailand; suranonmath@gmail.com

[2] Centre of Excellence in Mathematics, Commission on Higher Education, Ministry of Education, Si Ayutthaya Road, Bangkok 10400, Thailand

* Correspondence: ruthj.sku@mahidol.ac.th

Received: 2 September 2017; Accepted: 17 October 2017; Published: 23 October 2017

Abstract: The Radial Basis Function (RBF) method has been considered an important meshfree tool for numerical solutions of Partial Differential Equations (PDEs). For various situations, RBF with infinitely differentiable functions can provide accurate results and more flexibility in the geometry of computation domains than traditional methods such as finite difference and finite element methods. However, RBF does not suit large scale problems, and, therefore, a combination of RBF and the finite difference (RBF-FD) method was proposed because of its own strengths not only on feasibility and computational cost, but also on solution accuracy. In this study, we try the RBF-FD method on elliptic PDEs and study the effect of it on such equations with different shape parameters. Most importantly, we study the solution accuracy after additional ghost node strategy, preconditioning strategy, regularization strategy, and floating point arithmetic strategy. We have found more satisfactory accurate solutions in most situations than those from global RBF, except in the preconditioning and regularization strategies.

Keywords: numerical partial differential equations (PDEs); radial basis function-finite difference (RBF-FD) method; ghost node; preconditioning; regularization; floating point arithmetic

1. Introduction

For several decades, numerical solutions of partial differential equations (PDEs) have been studied by researchers in many different areas of science, engineering and mathematics. The common mathematical tools that are used to solve these problems are finite difference and finite element methods. These traditional numerical methods require the data to be arranged in a structured pattern and to be contained in a simply shaped region, and, therefore, these approaches does not suit multidimensional practical problems. The Radial basis function (RBF) method is viewed as an important alternative numerical technique for multidimensional problems because it contains more beneficial properties in high-order accuracy and flexibility on geometry of the computational domain than classical techniques [1–3]. The main idea of the RBF method is approximating the solution in terms of linear combination of infinitely differentiable RBF ϕ, which is the function that depends only on the distance to a center point $\underline{x}$ and shape parameter ε. More precisely, we approximate the solution as:

$$s(x,\varepsilon) = \sum_{j=1}^{N} \lambda_j \phi(\|x - x_j\|_2, \varepsilon),\tag{1}$$

where $x_j, j = 1, \ldots, N$ are the given centers. For simplicity, we set $\phi(r,\varepsilon) = \phi(\|x - \underline{x}\|_2, \varepsilon)$. Some common RBFs in the infinitely differentiable class are listed in Table 1.

Table 1. Radial basis functions.

Name of RBF	Abbreviation	Definition
Multiquadrics	MQ	$\phi(r, \varepsilon) = \sqrt{1 + (\varepsilon r)^2}$
Inverse Multiquadrics	IMQ	$\phi(r, \varepsilon) = \dfrac{1}{\sqrt{1 + (\varepsilon r)^2}}$
Inverse Quadratics	IQ	$\phi(r, \varepsilon) = \dfrac{1}{1 + (\varepsilon r)^2}$
Gaussians	GA	$\phi(r, \varepsilon) = e^{-(\varepsilon r)^2}$

For the case of interpolation of function $y = f(x)$, the coefficients $\{\lambda_j\}_{j=1}^N$ are determined by enforcing the following interpolation condition:

$$s(x_i, \varepsilon) = f(x_i) = y_i, \quad i = 1, 2, \ldots, N.$$

This condition results in an $N \times N$ linear system:

$$[A(\varepsilon)] \begin{bmatrix} \lambda_1 \\ \vdots \\ \lambda_N \end{bmatrix} = \begin{bmatrix} y_1 \\ \vdots \\ y_N \end{bmatrix}, \tag{2}$$

where the elements of $A(\varepsilon)$ are $a_{j,k} = \phi(\|x_j - x_k\|_2, \varepsilon)$. Note that the system matrix $A(\varepsilon)$ is known to be nonsingular if the constant shape parameter ε is used and is ill-conditioned for small values of shape parameter [4,5].

The concept of RBF interpolation can also be implemented for solving elliptic PDE problems. For the global RBF method, we assume the solution in the form of Equation (1). Then, the coefficients λ_j are determined by enforcing the PDE and boundary conditions. Unlike the case of interpolation, the system matrix of this case does not guarantee to be nonsingular and is also ill-conditioned for small values of shape parameter [1,6]. Additionally, according to numerical evidence, the global RBF scheme requires high computational cost and memory requirements for large scale problems [3,7]. Therefore, the radial basis function-finite difference (RBF-FD) concept, which is a local RBF scheme, was developed by combining many benefits of the RBF method and traditional finite difference approximations. Similar to the finite difference approach, the key idea is approximating the differential operator of solutions at each interior node by using a linear combination of the function values at the neighboring node locations and then determining the FD-weights so that the approximations become exact for all the RBFs that are centered at the neighboring nodes. Straightforward algebra will then show that the FD-weights can be obtained by solving the linear system, for which the system matrix is the same as matrix A in Equation (2). The invertibility of A also implies that FD-weights can always be computed. After the calculation at all interior nodes, the approximate solution can be computed from the linear system of equations, for which the RBF-FD system matrix is sparse and therefore can be effectively inverted. Note that the RBF-FD method that we described can be efficiently and applicably implemented with large scale practical problems such as the global electric circuit within the Earth's atmosphere problem [8], steady problems in solid and fluid mechanics [9–11], chemotaxis models [12], and diffusion problems [13,14].

In the first part of this work, the numerical solution of elliptic PDEs by using the RBF-FD method with IQ, MQ, IMQ, and GA RBF is compared to the results of the global RBF scheme. It will show that the accuracy of both global and local methods does not depend on the choice of RBFs, and the RBF-FD method with small stencils can obtain the same level of accuracy as the global RBF method. However, the RBF-FD method is algebraically accurate in exchange for low computational cost and needs more additional techniques to improve the accuracy [10,14]. Thus, in the latter part of this work, we will aim to research improving the accuracy of the RBF-FD method by using different ways to formulate the RBF-FD method. In particular, the effects of the number of nodes and stencils on the accuracy of numerical solutions of the RBF-FD method are investigated. According to [7,15], the global RBF method can improve the accuracy significantly for small values of shape parameter by using the regularization technique and extended precision floating point arithmetic to reduce the poor conditions, but how well these techniques can improve the accuracy of RBF-FD method is not yet fully explored in the literature. Therefore, the study of the RBF-FD method with these techniques will be included in this work as well.

2. Methodology

2.1. RBF Collocation Method

In this section, the global RBF scheme for solving the PDE problem called the RBF collocation method is introduced. Let $\Omega \subset \mathbb{R}^d$ be a d-dimensional domain and $\partial\Omega$ be the boundary of the domain. Consider the following elliptic PDE problem defined by:

$$\mathcal{L}u(x) = f(x) \quad \text{in } \Omega, \tag{3}$$

$$u(x) = g(x) \quad \text{on } \partial\Omega, \tag{4}$$

where $\mathcal{L}$ is a differential operator. Suppose that N is the number of center points in Ω, for which N_I points of them are the interior points. A straight RBF-based collocation method is applied by assuming the solution $u(x)$ as:

$$u(x) \approx \sum_{j=1}^{N} \lambda_j \phi(\|x - x_j\|_2, \varepsilon).$$

Consequently, collocation with the PDE at the interior points and the boundary condition at the boundary points provide the following results:

$$\mathcal{L}u(x_i) \approx \sum_{j=1}^{N} \lambda_j \mathcal{L}\phi(\|x_i - x_j\|_2, \varepsilon) = f(x_i), \quad i = 1, 2, \ldots, N_I,$$

$$u(x_i) \approx \sum_{j=1}^{N} \lambda_j \phi(\|x_i - x_j\|_2, \varepsilon) = g(x_i), \quad i = N_I + 1, N_I + 2, \ldots, N.$$

The coefficients $\{\lambda_j\}_{j=1}^{N}$ can be solved from the corresponding system of equations with the coefficient matrix structured as:

$$\begin{bmatrix} \mathcal{L}\phi \\ \phi \end{bmatrix} [\lambda] = \begin{bmatrix} f \\ g \end{bmatrix}.$$

2.2. RBF-FD Method

We will next present an outline of the RBF along with finite-difference (RBF-FD) formulation for solving PDE Equation (3).

First, let us look at a classical central finite difference method for approximating the derivative of function $u(x, y)$ with respect to x. Let the derivative at any grid point (i, j) of rectangular grid be written as

$$\left.\frac{\partial u}{\partial x}\right|_{(i,j)} \approx \sum_{k \in \{i-1, i, i+1\}} w_{(k,j)} u_{(k,j)}, \tag{5}$$

where $u_{(k,j)}$ is the function value at the grid point (k, j), the unknown coefficients $w_{(k,j)}$ are computed using polynomial interpolation or Taylor series, while the set of nodes $\{(i - 1, j), (i, j), (i + 1, j)\}$ is a stencil in the finite difference literature. For any scattered nodes, the restriction of the structured grid is overcome by setting the approximation of the function derivative to be a linear combination of function values on scattered nodes in the stencil; however, the methodology for computing the coefficients of finite difference formulas for any scattered points with dimensions of more than one has the problem of well-posedness for polynomial interpolation [16]. Thus, the combination of RBF and finite difference methodology (RBF-FD) is introduced to overcome the well-posedness problem.

To derive RBF-FD formulation, recall that, for a given set of distinct nodes $x_i \in \mathbb{R}^d, i = 1, 2, \ldots, n$, and a corresponding function values $u(x_i), i = 1, \ldots, n$, the RBF interpolation is of the form:

$$u(x) \approx \sum_{j=1}^{n} \lambda_j \phi(\|x - x_j\|_2, \varepsilon). \tag{6}$$

The RBF interpolation Equation (6) can alternatively be written in Lagrange form as:

$$u(x) \approx \sum_{j=1}^{n} \chi(\|x - x_j\|_2) u(x_j), \tag{7}$$

where $\chi(\|x - x_j\|_2)$ satisfies the following condition:

$$\chi(\|x - x_j\|_2) = \begin{cases} 1 & \text{if } x = x_j, \\ 0 & \text{if } x \neq x_j, \end{cases} \quad j = 1, \ldots, n.$$

The key idea of the RBF-FD method is to approximate the linear differential operator of function $\mathcal{L}u(x)$ at each interior node as a linear combination of the function values at the neighboring node locations. For example, $\mathcal{L}u(x)$ at any node, say x_1 is approximated by an RBF interpolation Equation (7) with centres placed on the node itself and some $n - 1$ nearest neighbor nodes of x_1, say $x_2, x_3, \ldots, x_n$. These set of n nodes is called the stencil or support region for the node x_1. For deriving RBF-FD formula at the node x_1, we approximate $\mathcal{L}u(x_1)$ by taking the linear differential operator $\mathcal{L}$ on the both sides of the RBF interpolation Equation (7) i.e.,

$$\mathcal{L}u(x_1) \approx \sum_{j=1}^{n} \mathcal{L}\chi(\|x_1 - x_j\|_2) u(x_j).$$

As in finite difference methodology, this approximation can be rewritten in the form:

$$\mathcal{L}u(x_1) \approx \sum_{j=1}^{n} w_{(1,j)} u(x_j), \tag{8}$$

where the RBF-FD weights $w_{(1,j)} = \mathcal{L}\chi(\|x_1 - x_j\|_2), j = 1, \ldots, n$.

In practice, the main difference between finite difference methodology and RBF-FD is how to compute the weights $w_{(1,j)}$ in Equation (8). While finite difference method enforces Equation (8) to be exact for polynomials $1, x, x^2, \ldots, x^{n-1}$, the RBF-FD weights are computed by assuming that the approximations Equation (8) become exact for all the RBFs ϕ that are centered at each node in stencil, i.e., we assume that Equation (8) becomes exact for function $u(x) = \phi(\|x - x_k\|_2, \varepsilon), k = 1, 2, \ldots, n$. In the RBF-FD case, this assumption leads to an $n \times n$ linear system:

$$\begin{bmatrix} \phi(\|x_1 - x_1\|_2, \varepsilon) & \phi(\|x_2 - x_1\|_2, \varepsilon) & \ldots & \phi(\|x_n - x_1\|_2, \varepsilon) \\ \phi(\|x_1 - x_2\|_2, \varepsilon) & \phi(\|x_2 - x_2\|_2, \varepsilon) & \ldots & \phi(\|x_n - x_2\|_2, \varepsilon) \\ \vdots & \vdots & \ddots & \vdots \\ \phi(\|x_1 - x_n\|_2, \varepsilon) & \phi(\|x_2 - x_n\|_2, \varepsilon) & \ldots & \phi(\|x_n - x_n\|_2, \varepsilon) \end{bmatrix} \begin{bmatrix} w_{(1,1)} \\ w_{(1,2)} \\ \vdots \\ w_{(1,n)} \end{bmatrix} = \begin{bmatrix} \mathcal{L}\phi(\|x_1 - x_1\|_2, \varepsilon) \\ \mathcal{L}\phi(\|x_1 - x_2\|_2, \varepsilon) \\ \vdots \\ \mathcal{L}\phi(\|x_1 - x_n\|_2, \varepsilon) \end{bmatrix}.$$

Note that the system matrix is invertible, and, therefore, the RBF-FD weight can always be calculated, while the system matrix for the finite difference case is not guaranteed to be invertible.

After we obtain RBF-FD weights, substituting the approximation Equation (8) into the elliptic PDE Equation (3) at node x_1 gives:

$$\sum_{j=1}^{n} w_{(1,j)} u(x_j) = f(x_1).$$

After that, we repeat this procedure at each of the interior nodes and substitute the approximations into the elliptic PDE Equation (3). Then, we obtain:

$$\sum_{j=1}^{n} w_{(i,j)} u(x_j) = f(x_i), \quad i = 1, 2, \ldots, N_I, \tag{9}$$

where N_I is the number of interior points. After substituting the function values $u(x_j)$ in Equation (9) with boundary condition whenever x_j is the boundary point, we obtain a linear system of equations, which can be written in matrix form as:

$$Bu = F, \tag{10}$$

where u is the vector of the unknown function at all the interior nodes. Note that the matrix B is sparse, well-conditioned and can hence be effectively inverted.

3. Main Result

3.1. Solution Accuracy for Elliptic PDEs

The main result section provides better understanding on the computational process. Let $\Omega \subset \mathbb{R}^2$ be a 2-dimensional domain and $\partial\Omega$ be the boundary of Ω. The elliptic PDE that we want to investigate is presented as the following:

$$\nabla^2 u(\underline{x}) + a(x,y)u_x(\underline{x}) + b(x,y)u_y(\underline{x}) + c(x,y)u(\underline{x}) = f(\underline{x}), \quad \underline{x} \text{ in } \Omega, \tag{11}$$
$$u(\underline{x}) = g(\underline{x}), \quad \underline{x} \text{ on } \partial\Omega. \tag{12}$$

In numerical experiments, the solution $u(\underline{x})$ is known and normalized so that $\max_{\Omega}|u(\underline{x})| = 1$. Supposing that the RBF approximation is denoted by $s(\underline{x}, \varepsilon)$, then the error in a max norm is measured by:

$$\text{Max Error}(\varepsilon) = \max_{\Omega}|s(\underline{x}, \varepsilon) - u(\underline{x})|. \tag{13}$$

3.2. Results: A Comparison of a Global RBF Collocation Method and RBF-FD Method with Various RBFs

In this section, the global RBF collocation method and RBF-FD method will be first investigated with several RBFs by applying them to the following elliptic PDE:

$$\nabla^2 u - Ru_x = f(x,y), \tag{14}$$

where $R = 3$, $f(x,y)$ and the Dirichlet boundary condition are computed from the exact solution as shown in Equation (15):

$$u(x,y) = \exp(-0.5Rx)\sin(0.5Ry). \tag{15}$$

For this problem, we consider the PDE with the computational and evaluational domain of a unit disk with 394 uniformly scattered data points as shown in Figure 1. In our first experiment, we compare the results from using the implementation of global RBF scheme and RBF-FD method with stencil size = 13. The results of the computation of the global RBF collocation method and RBF-FD method, which is the local RBF scheme, are shown in Figure 2.

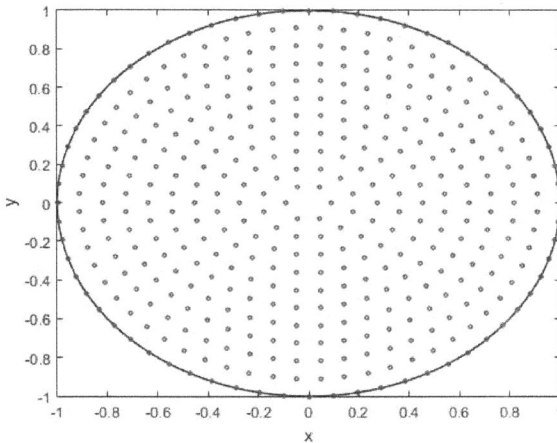

Figure 1. The computational and evaluational domain of a unit disk with 394 uniformly scattered data points.

Figure 2a shows the results of the computation of RBF collocation using IQ, IMQ, MQ, and GA radial basis functions, and Figure 2b shows the results of the computation of the RBF-FD method. Considering the global RBF scheme, all four of RBF seem to provide the same level of accuracy and fluctuated graph of max error for small values of shape parameter, while there is a little bit of difference in accuracy and smooth curve of the graph for the large values of shape parameter. The IQ, IMQ and MQ RBFs have similar behaviour of error's curves; however, the GA radial basis function has a different behaviour of error and gives the least error over the considered range of shape parameter for this problem. Note that all of them are spectrally accurate and the graph of error is oscillated since the small shape parameter makes the system matrix of global RBF collocation ill-conditioned, and, therefore, the numerical solution is computed inaccurately. This ill-conditioning problem will be more severe if more nodes are used. Due to this fact, the global RBF-scheme does not suit large scale problems.

For the RBF-FD case, all four of RBFs seem to provide the same level of accuracy for all shape parameters in the considered range. In particular, all of them have similar behaviour of error graphs and algebraic accuracy. For the large values of shape parameter, there is no significant differences of error between global and local RBF methods; however, the error of the global RBF scheme is more accurate by about two digits of accuracy than the RBF-FD method for the small values of shape parameter. Note that the graph of the error for the RBF-FD method does not oscillate for the small shape parameter because the system matrix at each node has better conditioning than the global RBF scheme for small stencil size. This fact makes the RBF-FD approach more appropriate for PDE problems with a large number of nodes.

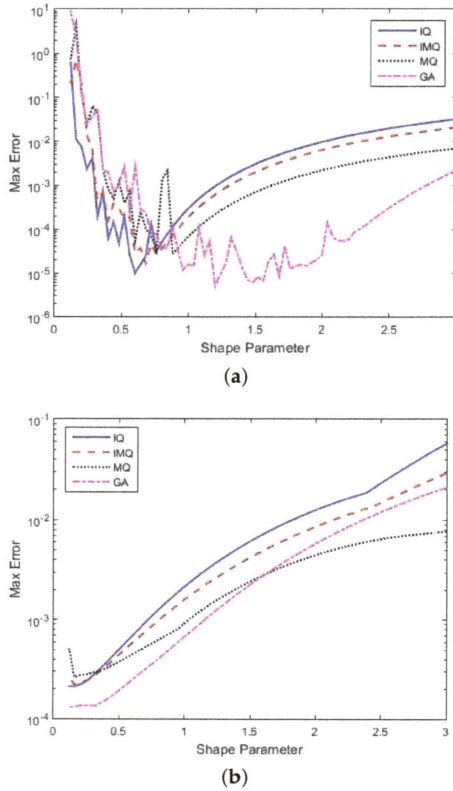

Figure 2. The graphs of max error are presented as a function of shape parameter. They compare max error of the results of global radial basis function (RBF) and radial basis function-finite difference (RBF-FD) methods. (**a**) global RBF scheme; (**b**) local RBF scheme.

3.3. Results: A Comparison of the RBF-FD Method with Different Numbers of Nodes and Stencils

From the previous section, the results show that the accuracy of RBF-FD is less than the accuracy of the global RBF scheme for the same number of nodes. However, the RBF-FD method is applicable and efficient for large scale problems. We hope that the increasing of the number of nodes will also help improve the accuracy for the case of the global RBF method, with the conditions for system matrix rapidly increased as well for the case of the RBF scheme [7,17]. Additionally, we expect that the accuracy will be more accurate whenever the stencil size increases. To ascertain this idea, we will consider the elliptic PDE given by:

$$\nabla^2 u - R(u_x + u_y) = f(x, y). \tag{16}$$

In this case, $R = 2$, $f(x, y)$ and Dirichlet boundary condition are computed from the exact solution as shown in Equation (17):

$$u(x, y) = \frac{(e^{Rxy} - 1)}{(e^R - 1)}. \tag{17}$$

For numerical experiments, we consider this problem in the computational and evaluational domain of a unit disk with uniformly scattered points. In our first experiment, the number of nodes are varied while the stencil size is fixed equal to 25, and then the effect of the number of nodes on the accuracy is compared. As for the second, the experiment is done in the opposite direction. The stencil sizes are varied while the number of nodes is fixed equal to 394, and then the effect of the stencil sizes on the accuracy is compared.

The results of computation using IQ-RBF when the number of nodes are varied and the stencil size is fixed are shown in Figure 3a. All of the results seem to provide the same level of accuracy for large values of shape parameter, while there are improvements on accuracy for the small shape parameter. These results provide a good simple way for improving accuracy. However, we need to be careful lest the ill-conditioning problem would arise if the stencil size is too large. For the case of varying the stencil sizes and the number of nodes being fixed as shown in Figure 3b, the results show that the more nodes used in the computation, the more accurate results can be achieved regardless of the shape parameter values. These results are also similar to the case of the global RBF method. In practical problems, this strategy of improving accuracy is very effective for the RBF-FD method with small enough stencil size since the sparse RBF-FD system matrix is well-conditioned and can be effectively inverted for a large number of nodes. However, this strategy does not suit a global RBF approach because the condition number of the system matrix will grow rapidly as the number of nodes increases and this will make the numerical solution inaccurate.

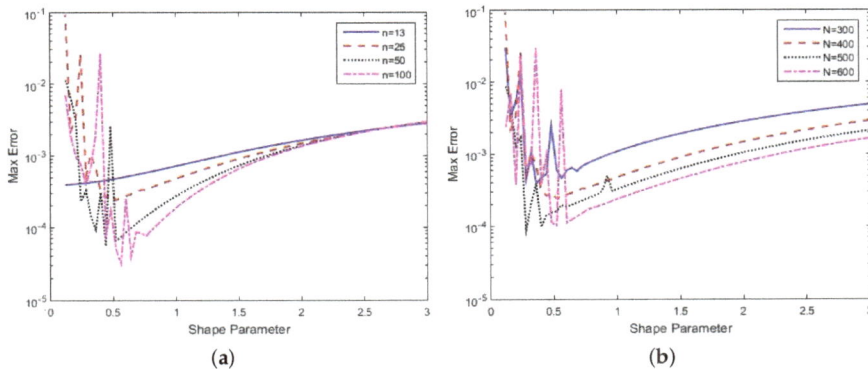

Figure 3. The graph of the max error of the RBF-FD method with different numbers of nodes and stencils. (**a**) varying number of stencils (n); (**b**) varying number of nodes (N).

3.4. Results: A Comparison of the RBF-FD Method with Strategies on PDE and Shape Parameter

In this section, the RBF-FD method with strategies on PDE and shape parameter are compared to the normal RBF-FD scheme. The elliptic PDE that we consider here is given by

$$\nabla^2 u + \exp(-(x+y))u = f(x,y), \tag{18}$$

where $f(x,y)$ and the Dirichlet boundary condition are computed from the exact solution as shown in Equation (19):

$$u(x,y) = \sinh(x^2 + y^2). \tag{19}$$

In this problem, the computations are implemented with the 395 uniformly scattered points and stencil size equal to 50.

- **Method 1 : Normal RBF-FD scheme**

As we discussed in Section 2.2, a straightforward RBF-FD method can be implemented for approximating the differential operator $\mathcal{L}$ by

$$\mathcal{L}u(x_j) \approx \sum_{i=1}^{n} w_{(j,i)} u(x_i),$$

for each interior node $x_j, j = 1, 2, \ldots, N$. After obtaining the FD-weights, we substitute the function values $u(x_j)$ from the boundary condition, and then a system of equations as shown in Equation (10) can be formed which can be written in matrix form as

$$Bu = F,$$

where u is the vector of the unknown function value at all of the interior nodes. Note that the matrix B is sparse and well-conditioned and can hence be effectively inverted.

- **Method 2 : RBF-FD method with ghost nodes strategy**

One of the simple ways to improve accuracy of numerical solutions is applying the PDE on boundary nodes. According to the global RBF method, the error seems to be largest near the boundary of the computational domain. This problem can be fixed by getting more information there by applying the PDE at the boundary nodes [1]. In order to match the number of unknowns and equations, additional nodes, called ghost nodes, are added. The centers of these ghost nodes should be placed outside the boundary. For simplicity, we let the center points be denoted by z_j, where

$$z_j = \begin{cases} x_j, & j = 1, 2, \ldots, N, \\ \text{a point outside } \Omega, & j = N+1, N+2, \ldots, N+N_B, \end{cases}$$

where N_B is the number of boundary nodes. After that, a straightforward RBF-FD scheme is implemented by approximating the differential operator $\mathcal{L}$ by

$$\mathcal{L}u(x_j) \approx \sum_{i=1}^{n} w_{(j,i)} u(x_i),$$

for all node $x_j, j = 1, 2, \ldots, N$. Note that the search for stencils must be applied to all nodes z_j. After obtaining the FD-weights and substituting the function values $u(x_j)$ from the boundary condition, a system of equations as shown in Equation (10) can be formed, which can be written in matrix form as

$$Bu = F,$$

where u is the vector of the unknown function value at all of the interior nodes and points outside the boundary. Note that the matrix B is sparse and well-conditioned and can hence be effectively inverted. In this problem, we use the additional 64 ghost nodes defined by $x_n = 1.1 \exp(\frac{2n\pi i}{64})$, $n = 1, 2, \ldots, 64$.

- **Method 3 : RBF-FD method with preconditioning strategy**

A variation that may improve the accuracy of the numerical solution is using the different values of shape parameter on each stencil. According to numerical evidence, RBF methods usually provide better accuracy if their system matrix is ill-conditioned. For computation with double precision, the shape parameter can be selected for each stencil so that the condition number is between 10^{13} and 10^{15} [12] because the error curve of the corresponding range of the shape parameter is the smooth curve before beginning to oscillate after that, as shown in Figure 2a. The following pseudocode will be

used to select the shape parameter in this problem so that the condition number is in the desired range:
K = 0, Kmin = 1.0 × 10¹³, Kmax = 1.0 × 10¹⁵, shape = initial shape, increment = 0.01

    while K < Kmin or K > Kmax :
        form B
        K = condition number of B
        if K < Kmin
            shape = shape - increment
        elseif K > Kmax
            shape = shape + increment.

Figure 4 shows the results of computation using IQ-RBF with strategies on PDE and shape parameter, compared to the normal RBF scheme. The ghost nodes strategy seems to provide the same level of accuracy for large values of shape parameter, while there is a significant improvement on accuracy for the small shape parameter. Note that using the additional ghost nodes strategy does not significantly affect the condition number of the system matrix. Therefore, for the practical problem, this approach can be implemented effectively to improve the error of numerical solution near the boundary. For the preconditioning strategy, the results show that, regardless of the initial shape parameter, this strategy can provide acceptable accurate results. Therefore, this approach is an applicable and effective strategy for choosing the shape parameter for the large scale practical problems.

After we choose the shape parameter, a normal RBF-FD method can be implemented for approximating the derivative and then forming a linear system for finding the approximate solution.

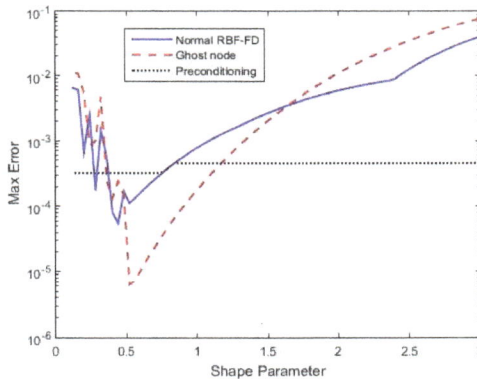

Figure 4. The graph shows the comparing of max error among the RBF-FD method with ghost nodes, the RBF-FD method with preconditioning, and the normal RBF-FD method. In the case of the RBF-FD method with preconditioning, we plot the max error against the initial shape parameter.

3.5. Results: A Comparison of RBF-FD Method with Additional Strategies for Reducing the Ill-Conditioning Problem

In the previous section, the RBF-FD method with ghost nodes and preconditioning strategies is applied for obtaining acceptable accurate results and reducing the accuracy of numerical solutions near the boundary. This section aims to improve the accuracy through another manipulation, i.e., by focusing on reducing the ill-conditioned problem for small values of shape parameters. To clarify this idea, we will investigate two alternative strategies called regularization and extended precision floating point arithmetic. The following elliptic PDE will be studied here :

$$\nabla^2 u + xu_x + yu_y - (4x^2 + y^2)u = f(x,y), \tag{20}$$

where $f(x, y)$ and the Dirichlet boundary condition are computed from the same solution as shown in Equation (21)

$$u(x, y) = \exp(-(x^2 + 0.5y^2)). \tag{21}$$

- **Strategy 1 : Regularization**

For each stencil, a simple regularization technique can be used to reduce the effects of the poor conditioning of the RBF system matrix by solving the regularized system

$$By = f \tag{22}$$

instead of the following system

$$Ax = f, \tag{23}$$

where $B = A + \mu I$. The regularization parameter μ is a small positive constant and I is an identity matrix. This approach can be used to solve effectively the ill-conditioning problem of the global RBF collocation method and give accurate results for small values of shape parameters, as shown in [15,18]. In this problem, we set $\mu = 5 \times 10^{-15}$.

- **Strategy 2 : Extended precision floating point arithmetic**

Recently, computer systems usually implement the IEEE 64-bit floating point arithmetic standard, which is referred to as double precision. According to [7], the global RBF method gains more benefit when it is implemented with extended precision, which provides more digits of accuracy than double precision. In particular, this approach also keeps the simplicity of the RBF method. In this work, the authors used the Multiprecision Computing Toolbox for Matlab (MCT) [19] to provide extended precision for the Matlab program and used quadruple precision (34 digits of precision) in this problem.

For numerical experiments, this problem is implemented with the computational and evaluational domain of a unit disk with 394 scattered data points and stencil size equal to 50. Figure 5 shows the results of computation using IQ-RBF with regularization and extended precision floating point arithmetic strategies, compared to the normal RBF-FD method. Unfortunately, the regularization seems to provide the same level of accuracy as the RBF-FD method without regularization, although this technique provides a significant improvement on accuracy compared to the global RBF method. However, for the extended precision strategy, the results show that this technique provides a significant improvement on accuracy at the same level as the global RBF method.

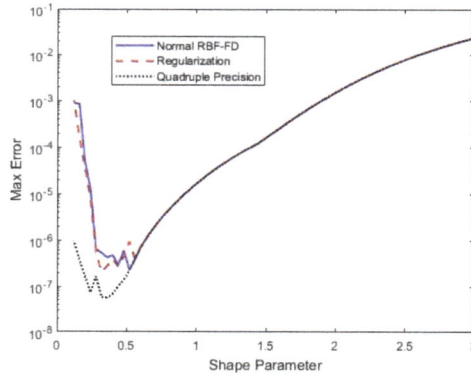

Figure 5. The graphs of max error of the RBF-FD method with regularization and extended precision strategies, compared to the normal RBF-FD method.

4. Future Work

RBF-FD method can also be easily extended to solve time-dependent PDEs by using the method of line such as the Runge–Kutta method (RK4) to solve the problem after the PDE is discretized in space with the RBF-FD method. For example, consider the diffusion problem given by

$$u_t = \nu \nabla^2 u + \gamma u^2 (1 - u), \tag{24}$$

where $\nu = 0.5, \gamma = 2$, with the Dirichlet boundary and initial condition is taken from the exact solution

$$u(x, y) = \frac{1}{1 + \exp(x + y - t)}. \tag{25}$$

To attack this problem, the Laplace operator ∇^2 is discretized by the RBF-FD method as we mentioned in Section 2.2. After we obtain the results, the remaining part will be ODE, the same as the following form

$$u_t = Du + \gamma u^2 (1 - u), \tag{26}$$

where D is the RBF-FD system matrix, which is obtained from discretization of the Laplace operator and u is the vector of the unknown function at all of the interior nodes. At this point, the fourth-order Runge–Kutta method (RK4) can be applied to solve the problem. The results that are implemented with the computational and evaluational domain of a unit disk with 394 scattered data points, stencil size equal to 13, and shape parameter equal to 0.6 are shown in Table 2.

Table 2. The results of solving the partial differential equation (PDE) Equation (24) with the radial basis function-finite difference (RBF-FD) method at $t = 5, 10, 15$.

t	5	10	15
Max Error	2.19×10^{-4}	2.57×10^{-4}	2.71×10^{-4}

Note that the RBF-FD method also gives reasonable numerical accuracy for time-dependent problems. However, many questions about eigenvalue stability of the RBF-FD method for time-dependent PDE problems in general have not yet been fully explored in the literature and need more study in future work.

5. Discussion

By using the RBF-FD method, we were able to solve large scale PDE problems due to the fact that RBF-FD has its own strengths related to feasibility and computational cost. We have found that accuracy can still be achieved at the same level of global RBF in the ill-conditioned system matrix for nodes. Although the RBF-FD method does not contain the same spectral accuracy as global RBF, it will give acceptable accuracy for large scale problems for which the global RBF can not be implemented. We also note that the accuracy can be improved by the increasing of nodes and stencil sizes. This strategy of improving accuracy is very effective with the RBF-FD method with small enough stencil size since the sparse RBF-FD system matrix is well-conditioned and can effectively be inverted for a large number of nodes. However, we need to be careful about an ill-conditioning problem arising if the stencil size is too large. The ghost node and preconditioning strategies are introduced to improve the accuracy of numerical solutions. Using the additional ghost node strategy does not significantly affect the condition number of the system matrix. Therefore, for the practical problem, this approach can be implemented effectively to reduce the errors of numerical solutions near boundaries. Results show that, regardless of the initial shape parameter, preconditioning strategy can provide acceptable accurate results. Improving the accuracy by reducing the ill-conditioning problem for small shape parameter is also an interesting idea. From experimental results, the extended precision provides significant improvement in accuracy. Unfortunately, regularization strategy fails to yield better accuracy than that of the plain RBF-FD.

Table 3 shows the results of solving PDE Equation (20) by using the RBF-FD method at a shape parameter equal to 0.4 with additional different strategies, which are presented in this work. In this case, preconditioning and regularization techniques yield the same level of accuracy as the normal RBF-FD method, while the other strategies can improve accuracy by around one digit of accuracy. Note that all of these strategies can be used together as a mixed strategy for obtaining high accuracy. However, the issues about how to choose the optimal shape parameter for obtaining the most accurate results and how to solve the ill-conditioned problem for small values of shape parameter are still open problems needing further investigation.

Table 3. The results of solving the PDE Equation (20) by using the RBF-FD method at a shape parameter equal to 0.4 with different additional strategies.

Method	Max Error ($\times 10^{-7}$)
Normal RBF-FD ($N = 400, n = 25$)	42.4
Normal RBF-FD ($N = 400, n = 50$)	4.75
Normal RBF-FD ($N = 800, n = 25$)	8.80
Ghost Node ($N = 400, n = 25$, 64 ghost nodes)	5.63
Preconditioning ($N = 400, n = 25$)	40.5
Regularization ($N = 400, n = 25$)	38.1
Extended Precision ($N = 400, n = 25$)	6.34

6. Conclusions

In this paper, we have found that the combination of the Radial Basis Function and the finite difference (RBF-FD) method with either additional Ghost Node strategy or additional Extended Precision strategy can help improve the accuracy in solving the elliptic partial differential equations, while the Preconditioning and Regularization strategies are found to be of little avail for our purpose.

Acknowledgments: This research was funded by the Faculty of Science, Mahidol University, Bangkok, Thailand.

Author Contributions: Suranon Yensiri and Ruth J. Skulkhu designed and developed the research. Suranon Yensiri was responsible for simulating the results by the MATLAB programming (version 9.1.0.441655 (R2016b), The MathWorks Inc., Natick, Massachusetts, United States). Both authors wrote this paper and contributed to editing and revising the manuscript.

Conflicts of Interest: The authors declare no conflict of interest.

References

1. Larsson, E.; Fornberg, B. A numerical study of some radial basis function based solution methods for elliptic PDEs. *Comput. Math. Appl.* **2003**, *46*, 891–902.
2. Li, J.; Cheng, H.-D.; Chen, C.-S. A comparison of efficiency and error convergence of multiquadric collocation method and finite element method. *Eng. Anal. Bound. Elem.* **2003**, *27*, 251–257.
3. Sarra, S.A.; Kansa, E.J. *Multiquadrics Radial Basis Function Approximation Methods for The Numerical Solution to Partial Differential Equations*; Tech Science Press: Duluth, GA, USA, 2010.
4. Fasshauer, G.E. *Meshfree Approximation Methods with Matlab*; World Scientific: Singapore, 2007.
5. Micchelli, C. Interpolation of scattered data: Distance matrices and conditionally positive definite functions. *Constr. Approx.* **1986**, *2*, 11–22.
6. Hon, Y.C.; Schaback, R. On unsymmetric collocation by radial basis functions. *Appl. Math. Comput.* **2001**, *119*, 177–186.
7. Sarra, S.A. Radial basis function approximation methods with extended precision floating point arithmetic. *Eng. Anal. Bound. Elem.* **2011**, *35*, 68–76.
8. Bayona, V.; Flyer, N.; Lucas, G.M.; Baumgaertner, A.J.G. A 3-D RBF-FD solver for modeling the atmospheric global electric circuit with topography (GEC-RBFFD v1.0). *Geosci. Model Dev.* **2015**, *8*, 3007–3020.

9. Shan, Y.; Shu, C.; Qin, N. Multiquadric finite difference (MQ-FD) method and its application. *Adv. Appl. Math. Mech.* **2009**, *5*, 615–638.
10. Tolstykh, A.I.; Lipavskii, M.V.; Shirobokov, D.A. High-accuracy discretization methods for solid mechanics. *Arch. Mech.* **2003**, *55*, 531–553.
11. Tolstykh, A.I.; Shirobokov, D.A. On using radial basis functions in finite difference mode with applications to elasticity problems. *Comput. Mech.* **2003**, *33*, 68–79.
12. Sarra, S.A. A Local Radial Basis Function method for Advection-Diffusion-Reaction equations on complexly shaped domains. *Appl. Math. Comput.* **2012**, *218*, 9853–9865.
13. Hosseini, B.; Hashemi, R. Solution of Burgers' equation using a local-RBF meshless method. *Int. J. Comput. Meth. Eng. Sci. Mech.* **2011**, *12*, 44–58.
14. Shankar, V.; Wright, G.B.; Fogelson, A.L.; Kirby, R.M. A radial basis function (RBF) finite difference method for the simulation of reaction-diffusion equations on stationary platelets within the augmented forcing method. *Int. J. Numer. Methods Fluids* **2014**, *75*, 1–22.
15. Sarra, S.A. Regularized symmetric positive definite matrix factorizations for linear systems arising from RBF interpolation and differentiation. *Eng. Anal. Bound. Elem.* **2014**, *44*, 76–86.
16. Wright, G.B.; Fornberg, B. Scattered node compact finite difference-type formulas generated from radial basis functions. *J. Comput. Phys.* **2006**, *212*, 99–123.
17. Sarra, S.A. A numerical study of the accuracy and stability of symmetric ans asymmetric RBF collocation methods for hyperbolic PDEs. *Numer. Methods Partial Differ. Equ.* **2008**, *24*, 670–686.
18. Sarra, S.A. The Matlab radial basis function toolbox. *J. Open Res. Softw.* **2017**, *5*, 8, doi:10.5334/jors.131.
19. Advanpix. Multiprecision Computing Toolbox for Matlab, Version 4.4.1.12566 for 64-bit Windows. Available online: http://www.advanpix.com (accessed on 17 August 2017).

![Σ mathematics logo] *mathematics*

MDPI

Article

Mixed Order Fractional Differential Equations

Michal Fečkan [1,2,*] and JinRong Wang [3]

[1] Department of Mathematical Analysis and Numerical Mathematics, Comenius University in Bratislava, Mlynská dolina, 842 48 Bratislava, Slovakia

[2] Mathematical Institute of Slovak Academy of Sciences, Štefánikova 49, 814 73 Bratislava, Slovakia

[3] Department of Mathematics, Guizhou University, Guiyang 550025, China; jrwang@gzu.edu.cn

* Correspondence: Michal.Feckan@fmph.uniba.sk; Tel.: +421-2-602-95-781

Received: 8 September 2017; Accepted: 31 October 2017; Published: 7 November 2017

Abstract: This paper studies fractional differential equations (FDEs) with mixed fractional derivatives. Existence, uniqueness, stability, and asymptotic results are derived.

Keywords: fractional differential equations (FDEs); Lyapunov exponent; stability

1. Introduction

Recently, fractional differential equations (FDEs) arise naturally in various fields, such as economics, engineering, and physics. For some existence results of FDEs we refer the reader to [1–6] and the references cited therein.

Bonilla et al. [1] studied linear systems of the same order linear FDEs and obtained an explicit representation of the solution. However, there are very few works on the study of mixed order nonlinear fractional differential equations (MOFDEs), which is a natural extension of [1].

This paper is devoted to the study of MOFDEs of the form

$$D_{0^+}^{q_i} x_i = f_i(t, x_1, \cdots, x_n), \quad x_i(0) = u_i, \quad i = 1, \cdots, n \tag{1}$$

where $D_{0^+}^{q_i}$ denotes the Caputo derivative with the lower limit at 0, $q_i \in (0, 1]$, $f : \mathbb{R}_+ \times \mathbb{R}^n \to \mathbb{R}^n$ are continuous, specified below and $u_i \in \mathbb{R}^n$. We may suppose $q_1 \geq \cdots \geq q_n$. Here $\mathbb{R}_+ = [0, \infty)$. We are interested in the existence of solutions of (1), then their stability and asymptotic properties under reasonable conditions on f_i. FDEs with equal order (i.e., $q_1 = \cdots = q_n$) are widely studied, and we refer the reader to the basic books describing FDEs, such as [7,8]. On the other hand, there are many MOFDEs with interesting applications—for example, to economic systems in [9]. In fact, (1) formulates a model of the national economies in a case of the study of n commonwealth countries, which cannot be simply divided into clear groups of independent and dependent variables. The purpose of this paper is to set a rigorous theoretical background for (1).

The main contributions are stated as follows:

We give some existence and uniqueness results for solutions of (1) when the nonlinear term satisfies global and local Lipschitz conditions.

We analyze the upper bound for Lyapunov exponents of solutions of (1).

We show that the zero solution of an autonomous version system of (1) is asymptotically stable.

2. Existence Results

First we prove an existence and uniqueness result for globally Lipschitzian $f = (f_1, \cdots, f_n)$. Let $\|x\|_\infty = \max_{i=1,\cdots,n} |x_i|$ for $x = (x_1, \cdots, x_n)$. By $C(J, \mathbb{R}^n)$ we denote the Banach space of all continuous functions from a compact interval $J \subset \mathbb{R}$ to $\mathbb{R}^n$ with the uniform convergence topology on J.

Theorem 1. *Let $T > 0$ and suppose the existence of $L > 0$ such that*

$$\|f(t,x) - f(t,y)\|_\infty \leq L\|x - y\|_\infty \quad \forall (t,x), (t,y) \in [0,T] \times \mathbb{R}^n. \tag{2}$$

Then (1) *has a unique solution* $x \in C(I, \mathbb{R}^n)$, $I = [0,T]$.

Proof of Theorem 1. Note that (2) implies

$$\|f(t,x)\|_\infty \leq L\|x\|_\infty + M_f \quad \forall (t,x) \in I \times \mathbb{R}^n \tag{3}$$

for $M_f = \max_{t \in I} \|f(t,0)\|_\infty$. Next, (1) is equivalent to the fixed point problem

$$x = F(x,u),$$
$$x(t) = (x_1(t), \cdots, x_n(t)), \quad u = (u_1, \cdots, u_n), \quad F(x,u) = (F_1(x,u), \cdots, F_n(x,u)),$$
$$F_i(x,u)(t) = u_i + \frac{1}{\Gamma(q_i)} \int_0^t (t-s)^{q_i-1} f_i(s, x_1(s), \cdots, x_n(s)) ds, \quad i = 1, \cdots, n. \tag{4}$$

Fix $\alpha \geq 0$ and set $\|x\|_\alpha = \max_{t \in [0,T]} \|x(t)\|_\infty e^{-\alpha t}$ for any $x \in C(I, \mathbb{R}^n)$. Let $x \in C(I, \mathbb{R}^n)$ be a solution of (1). For $\alpha > 0$, (3) and (4) give

$$|x_i(t)| \leq |u_i| + \frac{1}{\Gamma(q_i)} \int_0^t (t-s)^{q_i-1} \left(L\|x(s)\|_\infty + M_f \right) ds$$
$$\leq \|u\|_\infty + \frac{M_f T^{q_i}}{\Gamma(q_i+1)} + \frac{L\|x\|_\alpha}{\Gamma(q_i)} \int_0^t (t-s)^{q_i-1} e^{\alpha s} ds \leq \|u\|_\infty + \frac{M_f T^{q_i}}{\Gamma(q_i+1)} + \frac{L\|x\|_\alpha}{\alpha^{q_i}} e^{\alpha t},$$

which implies

$$\|x\|_\alpha \leq \|u\|_\infty + M_f \Theta + \frac{L\|x\|_\alpha}{\alpha^{q_m}}$$

for $\alpha > 1$ and

$$\Theta = \max_{i=1,\cdots,m} \frac{T^{q_i}}{\Gamma(q_i+1)}.$$

Hence

$$\|x\|_\alpha \leq \frac{\|u\|_\infty + M_f \Theta}{1 - \frac{L}{\alpha^{q_m}}},$$

for

$$\alpha > \max\{1, L^{1/q_m}\}. \tag{5}$$

So

$$|x_i(t)| \leq \frac{\|u\|_\infty + M_f \Theta}{1 - \frac{L}{\alpha^{q_m}}} e^{\alpha T}, \quad t \in I, \quad i = 1, \cdots, n. \tag{6}$$

Similarly, we derive

$$\|F(x,u) - F(y,u)\|_\alpha \leq \frac{L}{\alpha^{q_m}} \|x - y\|_\alpha, \quad \forall x, y \in C(I, \mathbb{R}^n).$$

Consequently, assuming (5), we can apply the Banach fixed point theorem to get a unique solution $x \in C(I, \mathbb{R}^n)$ of (1), which also satisfies (6). The proof is finished. $\square$

Now we prove an existence and uniqueness result for locally Lipschitzian f.

Theorem 2. *Suppose that for any $r > 0$ there is an $L_r > 0$ such that*

$$\|f(t,x) - f(t,y)\|_\infty \leq L_r \|x - y\|_\infty \quad \forall (t,x), (t,y) \in [0,T] \times \mathbb{R}^n, \quad \max\{\|x\|_\infty, \|y\|_\infty\} \leq r. \tag{7}$$

Then (1) *has a unique solution* $x \in C(I_0, \mathbb{R}^n)$, $I_0 = [0, T_0]$ *for some* $0 < T_0 \le T$.

Proof of Theorem 2. Set $r_0 = \|u\|_\infty + M_f \Theta + 1$ and $B(r_0) = \{x \in \mathbb{R}^n \mid \|x\|_\infty \le r_0\}$. Then we extend f from the set $I \times B(r_0)$ to $\tilde{f}$ on $I \times \mathbb{R}^n$ such that $\tilde{f}$ satisfies (2) for some $L > 0$. This extension is given by

$$\tilde{f}(t, x) = \begin{cases} \chi(\|x\|_\infty/(r_0 + 1)) f(t, x) & \text{for } t \in I, \|x\| \le 2(r_0 + 1) \\ 0 & \text{for } t \in I, \|x\| \ge 2(r_0 + 1) \end{cases}$$

for a Lipschitz function $\chi : \mathbb{R}_+ \to [0, 1]$ with $\chi(r) = 1$ for $r \in [0, 1]$ and $\chi(r) = 0$ for $r \ge 2$. Applying Theorem 1 to

$$D_{0^+}^{q_i} x_i = \tilde{f}_i(t, x_1, \cdots, x_n), \quad x_i(0) = u_i, \quad i = 1, \cdots, n, \tag{8}$$

there is a unique solution $x \in C(I, \mathbb{R}^n)$ of (8) which also satisfies (6) for α satisfying (5). Note $M_f = M_{\tilde{f}}$. Let us take $T \ge T_0 > 0$, $\alpha > 0$ satisfying (5) and

$$\frac{\|u\|_\infty + M_f \Theta}{1 - \frac{L}{\alpha^{qm}}} e^{\alpha T_0} < r_0.$$

Then the unique solution $x \in C(I, \mathbb{R}^n)$ of (8) satisfies

$$\|x(t)\|_\infty \le r_0 \quad \forall t \in I_0.$$

However, this is also a unique solution of (1) on I_0. The proof is finished. $\square$

Remark 1. *Let us denote by* x_u *the solution from Theorem* 1. *Then, following the proof of Theorem* 1, *for any* $u, v \in \mathbb{R}^n$, *we derive*

$$\|x_u - x_v\|_\alpha = \|F(x_u, u) - F(x_v, v)\|_\alpha \le \frac{L}{\alpha^{qm}} \|x_u - x_v\|_\alpha + \|F(x_v, u) - F(x_v, v)\|_\alpha$$

$$\le \frac{L}{\alpha^{qm}} \|x_u - x_v\|_\alpha + \|u - v\|_\infty,$$

which implies

$$\|x_u - x_v\|_\alpha \le \frac{\|u - v\|_\infty}{1 - \frac{L}{\alpha^{qm}}},$$

i.e.,

$$\|x_u(t) - x_v(t)\|_\infty \le \frac{\|u - v\|_\infty}{1 - \frac{L}{\alpha^{qm}}} \quad \forall t \in I,$$

provided that (5) *holds. So, the continuous dependence of the solution of* (1) *is shown on the initial value* u *under conditions of Theorem* 1 *or Theorem* 2.

3. Asymptotic Results

We find the upper bound for Lyapunov exponents of solutions of (1).

Theorem 3. *Suppose assumptions of Theorem* 2 *are satisfied. Moreover, we suppose the existence of a nonnegative* $n \times n$-matrix $M = \{m_{ij}\}_{i,j \ge 1}^n$ *and a nonegative vector* $v = (v_1, \cdots, v_n)$ *such that*

$$|f_i(t, x)| \le \sum_{j=1}^n m_{ij} |x_j| + v_i \quad \forall (t, x) \in \mathbb{R}_+ \times \mathbb{R}^n, \quad i = 1, \cdots, n. \tag{9}$$

Then the Lyapunov exponent

$$\Lambda(u) = \limsup_{t \to \infty} \frac{\ln \|x(t)\|_\infty}{t}$$

of the unique solution $x \in C(\mathbb{R}_+, \mathbb{R}^n)$ of (1) *satisfies*

$$\Lambda(u) \le \rho(M)^{\max_{i=1,\cdots,n} 1/q_i},$$

where $\rho(M)$ is the spectral radius of M. Note $x(0) = u$, so we consider as usually $\Lambda : \mathbb{R}^n \to \mathbb{R}_+$ (see [10]).

Proof. Clearly (9) implies

$$\|f(t,x)\|_\infty \le \left(\max_{i=1,\cdots,n} \sum_{j=1}^{n} m_{ij} \right) \|x\|_\infty + \|v\|_\infty \quad \forall (t,x) \in \mathbb{R}_+ \times \mathbb{R}^n.$$

Then, like in the proof of Theorem 1, for any $T > 0$ there is a unique solution of (1) on I. However, since $T > 0$ is arbitrary, we get a unique solution $x(t)$ on $\mathbb{R}_+$. Then, we compute

$$|x_i(t)| \le |u_i| + \frac{1}{\Gamma(q_i)} \int_0^t (t-s)^{q_i-1} \sum_{j=1}^{n} \left(m_{ij}|x_j(s)| + v_j \right) ds$$

$$= |u_i| + \frac{t^{q_i}}{\Gamma(q_i+1)} \sum_{j=1}^{n} m_{ij} v_j + \frac{1}{\Gamma(q_i)} \sum_{j=1}^{n} m_{ij} \int_0^t (t-s)^{q_i-1} e^{-\alpha s} |x_j(s)| e^{\alpha s} ds$$

$$\le |u_i| + \frac{t^{q_i}}{\Gamma(q_i+1)} \sum_{j=1}^{n} m_{ij} v_j + \frac{1}{\Gamma(q_i)} \sum_{j=1}^{n} m_{ij} \|x_j\|_\alpha \frac{\Gamma(q_i) e^{\alpha t}}{\alpha^{q_i}}$$

$$= |u_i| + \frac{t^{q_i}}{\Gamma(q_i+1)} \sum_{j=1}^{n} m_{ij} v_j + \frac{e^{\alpha t}}{\alpha^{q_i}} \sum_{j=1}^{n} m_{ij} \|x_j\|_\alpha$$

$$\le |u_i| + Y_\alpha e^{\alpha t} \sum_{j=1}^{n} m_{ij} v_j + \frac{e^{\alpha t}}{\alpha^{q_i}} \sum_{j=1}^{n} m_{ij} \|x_j\|_\alpha$$

for

$$Y_\alpha = \max_{i=1,\cdots,n, t \in \mathbb{R}_+} \frac{t^{q_i} e^{-\alpha t}}{\Gamma(q_i+1)} = \max_{i=1,\cdots,n} \frac{q_i^{q_i} e^{-q_i}}{\alpha^{q_i} \Gamma(q_i+1)}.$$

Consequently, we arrive at

$$\|x_i\|_\alpha \le |u_i| + Y_\alpha \sum_{j=1}^{n} m_{ij} v_j + \frac{1}{\alpha^{q_i}} \sum_{j=1}^{n} m_{ij} \|x_j\|_\alpha,$$

or setting $w_\alpha = (\|x_1\|_\alpha, \cdots, \|x_m\|_\alpha)$ and $|u| = (|u_1|, \cdots, |u_n|)$, we obtain

$$w_{\alpha_\varepsilon} \le |u| + Y_{\alpha_\varepsilon} M v + \frac{1}{\rho(M) + \varepsilon} M w_{\alpha_\varepsilon}, \tag{10}$$

for $\alpha_\varepsilon = \max_{i=1,\cdots,n} (\rho(M) + \varepsilon)^{1/q_i}$ and $\varepsilon > 0$ is fixed. (10) is considered component-wise. From (10), we get

$$\left(I - \frac{1}{\rho(M) + \varepsilon} M \right) w_{\alpha_\varepsilon} \le |u| + Y_{\alpha_\varepsilon} M v \tag{11}$$

for the $n \times n$ identity matrix I. By the Neumann lemma, we have

$$\left(I - \frac{1}{\rho(M) + \varepsilon} M \right)^{-1} = \sum_{i=0}^{\infty} \left(\frac{1}{\rho(M) + \varepsilon} M \right)^i,$$

which is a positive matrix. Consequently, (11) implies

$$w_{\alpha_\varepsilon} \leq \left(I - \frac{1}{\rho(M) + \varepsilon}M\right)^{-1}(|u| + Y_{\alpha_\varepsilon}Mv).$$

Letting $\varepsilon \to 0$, we get

$$w_{\alpha_0} \leq \left(I - \frac{1}{\rho(M)}M\right)^{-1}(|u| + Y_{\alpha_0}Mv)$$

for $\alpha_0 = \max_{i=1,\cdots,n}\rho(M)^{1/q_i}$. So we arrive at

$$|x_i(t)| \leq e^{\alpha_0 t}w_i \quad \forall t \in \mathbb{R}_+, \quad i = 1,\cdots,n$$

for $w = (w_1, \cdots, w_n) = \left(I - \frac{1}{\rho(M)}M\right)^{-1}(|u| + Y_{\alpha_0}Mv)$, which implies

$$\Lambda(u) = \limsup_{t\to\infty}\frac{\ln\|x(t)\|_\infty}{t} \leq \limsup_{t\to\infty}\frac{\ln(e^{\alpha_0 t}\|w\|_\infty)}{t}$$

$$= \limsup_{t\to\infty}\frac{\ln e^{\alpha_0 t}}{t} + \limsup_{t\to\infty}\frac{\|w\|_\infty}{t} = \alpha_0.$$

The proof is finished. □

4. Stability Result

This section is devoted to an autonomous version of (1)

$$D_{0^+}^{q_i}x_i = \sum_{j=1}^n a_{ij}x_j + g_i(x_1,\cdots,x_n), \quad x_i(0) = u_i, \quad i = 1,\cdots,n \tag{12}$$

where $a_{i,j} \in \mathbb{R}$, $g = (g_1,\cdots,g_n) : \mathbb{R}^n \to \mathbb{R}^n$ is locally Lipschitz with $g(0) = 0$ and $\lim_{x\to 0}\frac{\|g(x)\|_\infty}{\|x\|_\infty} = 0$. We already know that (12) has a locally unique solution. First, we prove the following

Lemma 1. *Let $\lambda > 0$ and $q, p \in (0,1]$ with $q \geq p$. Then, it holds*

$$S(\lambda, q, p) = \sup_{t\geq 0}(t+1)^p \int_0^t (t-s)^{q-1}E_{q,q}(-(t-s)^q\lambda)(s+1)^{-p}ds < \infty. \tag{13}$$

Note that $E_{q,q}(-t) \geq 0$ for any $t \in \mathbb{R}_+$ by ([11] p. 85), where $E_{q,q}(t)$ is the two-parametric Mittag-Leffler function ([11] p. 56).

Proof. By ([5] Proposition 2.4(i))or ([11] Formula (4.4.17)), there is a positive constant $M(q,\lambda)$ such that

$$t^{q-1}E_{q,q}(-t^q\lambda) \leq \frac{M(q,\lambda)}{(t+1)^{q+1}} \quad \forall t \geq 1.$$

Here we note that one can use ([6] Lemma 3) to compute $M(q,\lambda)$. First, using ([11] Formula (4.4.4)), we derive

$$\sup_{t\in[0,1]}(t+1)^p \int_0^t (t-s)^{q-1}E_{q,q}(-(t-s)^q\lambda)(s+1)^{-p}ds$$

$$\leq \sup_{t\in[0,1]}(t+1)^p \int_0^t (t-s)^{q-1}E_{q,q}(-(t-s)^q\lambda)ds$$

$$= \sup_{t\in[0,1]}(t+1)^p \int_0^t z^{q-1}E_{q,q}(-z^q\lambda)dz = \sup_{t\in[0,1]}(t+1)^p t^q E_{q,q+1}(-\lambda t^q) = 2^p E_{q,q+1}(-\lambda),$$

since applying ([11] Formula (5.1.15)), we derive

$$\frac{d}{dt}(t+1)^p t^q E_{q,q+1}(-\lambda t^q) = p(t+1)^{p-1} t^q E_{q,q+1}(-\lambda t^q) + (t+1)^p t^{q-1} E_{q,q}(-\lambda t^q) \geq 0$$

for $t > 0$. Next, for $t \geq 1$, we derive

$$\sup_{t \geq 1}(t+1)^p \int_0^t (t-s)^{q-1} E_{q,q}(-(t-s)^q \lambda)(s+1)^{-p} ds$$

$$\leq \sup_{t \geq 1}(t+1)^p \int_0^{t-1} (t-s)^{q-1} E_{q,q}(-(t-s)^q \lambda)(s+1)^{-p} ds$$

$$+ \sup_{t \geq 1}(t+1)^p \int_{t-1}^t (t-s)^{q-1} E_{q,q}(-(t-s)^q \lambda)(s+1)^{-p} ds$$

$$\leq \sup_{t \geq 1}(t+1)^p \int_0^{t-1} \frac{M(q,\lambda)}{(t-s+1)^{q+1}(s+1)^p} ds + \sup_{t \geq 1} \frac{(t+1)^p}{t^p} \int_0^1 z^{q-1} E_{q,q}(-z^q \lambda) dz$$

$$\leq \sup_{t \geq 1}(t+1)^p \int_0^{t/2} \frac{M(q,\lambda)}{(t-s+1)^{q+1}} ds + \sup_{t \geq 1} \frac{2^p(t+1)^p}{(t+2)^p} \int_{t/2}^t \frac{M(q,\lambda)}{(t-s+1)^{q+1}} ds + 2^p E_{q,q+1}(-\lambda)$$

$$\leq \sup_{t \geq 1} \frac{2^q(t+1)^p M(q,\lambda)}{q(t+2)^q} + \sup_{t \geq 1} \frac{2^p(t+1)^p M(q,\lambda)}{q(t+2)^p} + 2^p E_{q,q+1}(-\lambda)$$

$$\leq \frac{(2^q + 2^p)M(q,\lambda)}{q} + 2^p E_{q,q+1}(-\lambda)$$

Summarizing, we arrive at the estimate

$$S(\lambda, q, p) \leq \frac{(2^q + 2^p)M(q,\lambda)}{q} + 2^p E_{q,q+1}(-\lambda).$$

The proof is finished. □

Theorem 4. *Suppose* $a_{ii} = -\lambda_i < 0, i = 1, \cdots, n$. *If*

$$\gamma = \max_{i=1,\cdots,n} S(\lambda_i, q_i, q_n) \sum_{i \neq j=1}^n |a_{ij}| < 1, \tag{14}$$

then the zero solution of (12) *is asymptotically stable.*

Proof. The proof is motivated by the well-known Geršgoring type method [12,13]. Like above, we modify (12) outside of the unit ball $B(1)$ such that the modified system is globally Lipschitz. Then, the solution of the modified (12) has a global unique solution on $\mathbb{R}_+$ by Theorem 1. This solution of (12) has the fixed point form [8]

$$x_i(t) = G_i(x, u), \quad i = 1, \cdots, n,$$

$$G_i(x, u) = E_{q_i}(-t^{q_i}\lambda_i)u_i + \int_0^t (t-s)^{q_i-1} E_{q_i,q_i}(-(t-s)^{q_i}\lambda_i) \left(\sum_{i \neq j=1}^n a_{ij}x_j(s) + g_i(x_1(s), \cdots, x_n(s)) \right) ds, \tag{15}$$

where $E_q(t)$ and $E_{q,q}(t)$ are the classical and two-parametric Mittag-Leffler functions, respectively ([11] p. 56). Fix $T > 0$ and set $|||x||| = \max_{t \in I} |x(t)|_\infty (t+1)^{q_n}$ for $x \in C(I, \mathbb{R}^n)$. Then, (15) implies

$$|G_i(x,u)(t)|(t+1)^{q_n} \leq \kappa|u_i| + (t+1)^{q_n} \int_0^t (t-s)^{q_i-1} E_{q_i,q_i}(-(t-s)^{q_i}\lambda_i)(s+1)^{-q_n} ds$$

$$\times \left(\sum_{\substack{i \neq j=1}}^n |a_{ij}| + h(\|x(s)\|_\infty) \right)|||x|||)$$

$$\leq \kappa\|u\|_\infty + S(\lambda_i,q_i,q_n)\left(\sum_{\substack{i \neq j=1}}^n |a_{ij}| + h(|||x|||)\right)|||x|||)$$

for

$$\kappa = \max_{i=1,\cdots,n} \sup_{t\geq 0} E_{q_i}(-t^{q_i}\lambda_i)(t+1)^{q_n} < \infty$$

(see ([11] Formula (3.4.30)) and $h(r) = \max_{\|x\|_\infty \leq r} \frac{\|g(x)\|_\infty}{\|x\|_\infty}$. Consequently, we obtain

$$|||G(x,u)||| \leq \kappa\|u\|_\infty + (\gamma + \omega h(|||x|||))|||x||| \tag{16}$$

for $G(x,u) = (G_1(x,u),\cdots,G_n(x,u))$ and

$$\omega = \max_{i=1,\cdots,n} \sum_{\substack{i \neq j=1}}^n S(\lambda_i,q_i,q_n).$$

Taking $r_1 \in (0,1)$ such that $\gamma + \omega h(r_1) \leq \frac{1+\gamma}{2}$, then (16) implies

$$|||G(x,u)||| \leq \kappa\|u\|_\infty + \frac{1+\gamma}{2}|||x||| \tag{17}$$

for any $x \in B_T(r_1) = \{x \in C(I,\mathbb{R}^n) \mid |||x||| \leq r_1\}$. So, supposing

$$\|u\| \leq \frac{1-\gamma}{2\kappa}r_1, \tag{18}$$

(17) gives

$$G : B_T(r_1) \subset B_T(r_1).$$

It is well-known that $G : C(I,\mathbb{R}^n) \to C(I,\mathbb{R}^n)$ is continuous and compact [8]. Since $B_T(r_1)$ is convex and bounded, by the Schauder fixed point theorem, G has a fixed point $x \in B_T(r_1)$. However, this a solution of a modified (12), which has a unique solution on $\mathbb{R}_+$. So, this unique solution satisfies $x \in B_T(r_1)$ for any $t > 0$; i.e.,

$$\|x(t)\|_\infty \leq \frac{r_1}{(t+1)^{q_n}} \quad \forall t \geq 0. \tag{19}$$

Certainly (19) gives $\|x(t)\|_\infty \leq 1$, so $x(t)$ is also a unique solution of the original (12). Moreover, (17) leads to

$$\|x(t)\|_\infty \leq \frac{2\kappa\|u\|_\infty}{(1-\gamma)(t+1)^{q_n}} \quad \forall t \geq 0,$$

which determines the asymptotic stability of (12) at 0. The proof is finished. $\square$

Remark 2. *Condition (14) is a Geršgoring type assumption [12,13]. Even for $q_1 = \cdots = q_n$, condition (14) is new, since the general stability assumption (see [5] Theorem 2.2) is difficult to check. The above approach can be extended to FDEs on infinite lattices like in [14].*

5. Examples

In this section, we give examples to demonstrate the validity of our theoretical results.

Example 1. *Consider*

$$\begin{cases} {}^cD_{0^+}^{0.6}x_1 = \sin(x_1 + 2x_2), \ t \in [0,1], \\ {}^cD_{0^+}^{0.5}x_2 = \arctan(2x_1 + x_2), \\ x_1(0) = u_1, \quad x_2(0) = u_2. \end{cases} \tag{20}$$

Set $q_1 = 0.6 > 0.5 = q_2$, $T = 1$, $f_1(t, x_1, x_2) = \sin(x_1 + 2x_2)$ and $f_2(t, x_1, x_2) = \arctan(2x_1 + x_2)$. *Clearly,* $f = (f_1, f_2)$ *satisfying the uniformly Lipschitz condition with a Lipschitz constant* $L = 2$. *By Theorem 1 or 2, (20) has a unique solution* $x \in C([0,1], R^2)$.
Next, $|f_1(t, x_1, x_2)| \leq m_{11}|x_1| + m_{12}|x_2|$ *and* $|f_2(t, x_1, x_2)| \leq m_{21}|x_1| + m_{22}|x_2|$ *where* $m_{11} = 1 = m_{22}$ *and* $m_{12} = m_{21} = 2$ *Set* $M = (m_{ij}) \in \mathbb{R}^{2 \times 2}$, *so*

$$M = \begin{pmatrix} 1 & 2 \\ 2 & 1 \end{pmatrix},$$

is a nonnegative matrix and $\rho(M) = 3$. *By Theorem 3,* $\Lambda(u) = 3^{\frac{5}{3}}$.

Example 2. *Consider*

$$\begin{cases} {}^cD_{0^+}^{0.6}x_1 = -x_1 - 0.5x_2 + \frac{x_1 + x_2}{1 + (x_1 + x_2)^2}\sin(x_1 + x_2), \ t \geq 0, \\ {}^cD_{0^+}^{0.5}x_2 = 0.5x_1 - x_2 + \frac{x_1 + x_2}{1 + (x_1 + x_2)^2}\arctan(x_1 + x_2). \end{cases} \tag{21}$$

Set $q_1 = 0.6 > 0.5 = q_2$, $\lambda_1 = \lambda_2 = 1$, $a_{11} = a_{22} = -1$ *and* $a_{12} = -0.5, a_{21} = 0.5$, $g_1(x_1, x_2) = \frac{x_1 + x_2}{1 + (x_1 + x_2)^2}\sin(x_1 + x_2)$, $g_2(x_1, x_2) = \frac{x_1 + x_2}{1 + (x_1 + x_2)^2}\arctan(x_1 + x_2)$ *and* $u_1 = u_2 = 0$. *Clearly,* $g = (g_1, g_2)$ *satisfies the uniformly Lipschitz condition with a Lipschitz constant* $L = 2$ *and* $g((0,0)) = (0,0)$ *and* $\lim_{(x_1, x_2) \to (0,0)} \frac{\|g(x_1, x_2)\|_\infty}{\|(x_1, x_2)\|_\infty} = 0$. *We numerically derive*

$$S(1, 0.6, 05) \doteq 1.0051, \quad S(1, 0.5, 0, 5) \doteq 1$$

so $\gamma \leq 0.503 < 1$. *By Theorem 4, the zero solution of (21) is asymptotically stable.*

6. Conclusions

The existence and uniqueness of solutions of (1) are shown along with their stability and asymptotic properties. Theorems 1 and 2 extended the partial case of $q_i = q, i = 1, 2, \cdots, n$ which are known in the related literature. Theorems 3 and 4 are original results. Moreover, Theorems 1, 2, and 3 can be directly extended to infinite dimensional cases. The next step should be to derive an exact solution for scalar linear systems (i.e., (12) with $g = 0$) and to find an explicit variation of constant formula for nonhomogeneous systems. Of course, a more general stability criterion would also be interesting to find by generalizing ([5] Theorem 2.2), which is just for $q_1 = \cdots = q_n$. Then, a derivation of a Gronwall-type inequality associated to MOFDEs would be challenging as well. This should extend Theorem 4 to infinite dimensional cases.

Acknowledgments: The authors thanks the referees for their careful reading of the manuscript and insightful comments, which help to improve the quality of the paper. This work is supported by National Natural Science Foundation of China (11661016) and the Slovak Research and Development Agency under the contract No. APVV-14-0378 and by the Slovak Grant Agency VEGA No. 2/0153/16 and No. 1/0078/17.

Author Contributions: M.F. and J.R.W. contributed equally in this work.

Conflicts of Interest: The authors declare no conflict of interest.

References

1. Bonilla, B.; Rivero, M.; Trujillo, J.J. On systems of linear fractional differential equations with constant coefficients. *Appl. Math. Comput.* **2007**, *187*, 68–78.
2. Chalishajar, D.N.; Karthikeyan, K.; Trujillo, J.J. Existence of mild solutions for fractional impulsive semilinear integro-differential equations in Banach spaces. *Commun. Appl. Nonlinear Anal.* **2012**, *4*, 45–56.
3. Chalishajar, D.N.; Karthikeyan, K. Existence and uniqueness results for boundary value problems of higher order fractional integro-differential equations involving Gronwall's inequality in banach spaces. *Acta Math. Sci.* **2013**, *33*, 758–772, doi:10.1016/S0252-9602(13)60036-3.
4. Chalishajar, D.N.; Karthikeyan, K. Boundary value problems for impulsive fractional evolution integro-differential equations with Gronwall's inequality in Banach spaces. *Discontin. Nonlinearity Complex.* **2014**, *3*, 33–48.
5. Cong, N.D.; Doan, T.S.; Siegmund, S.; Tuan, H.T. Linearized asymptotic stability for fractional differential equations. *Electron. J. Qual. Theory Differ. Equ.* **2016**, *39*, 1–13, doi:10.14232/ejqtde.2016.1.39.
6. Cong, N.D.; Doan, T.S.; Siegmund, S.; Tuan, H.T. On stable manifolds for planar fractional differential equations. *Appl. Math. Comput.* **2014**, *226*, 157–168, doi:10.1016/j.amc.2013.10.010.
7. Kilbas, A.A.; Srivastava, H.M.; Trujillo, J.J. *Theory and Applications of Fractional Differential Equations*; Elsevier: Amsterdam, The Netherlands, 2006.
8. Zhou, Y. *Fractional Evolution Equations and Inclusions: Analysis and Control*; Academic Press: Amsterdam, The Netherlands, 2016.
9. Škovránek, T, Podlubny, I., Petráš, I. Modeling of the national economies in state-space: A fractional calculus approach. *Econ. Model.* **2012**, *29*, 1322–1327, doi:10.1016/j.econmod.2012.03.019.
10. Barreira, L.; Pesin, Y. *Lectures on Lyapunov Exponents and Smooth Ergodic Theory*; American Mathematical Society: Providence, RI, USA, 1998.
11. Gorenflo, R.; Kilbas, A.A.; Mainardi, F.; Rogosin, S.V. *Mittag-Leffler Functions, Related Topics and Applications*; Springer: Berlin, Germany, 2014.
12. Gil, M.I. *Operator Functions and Localization of Spectra*; Springer: Berlin, Germany, 2003.
13. Varga, R.S. *Geršgorin and His Circles*; Springer: Berlin, Germany, 2011.
14. Fečkan, M.; Kelemen, S. Multivalued integral manifolds in Banach spaces. *Commun. Math. Anal.* **2011**, *10*, 97–117.

mathematics

MDPI

Article

Solution of Inhomogeneous Differential Equations with Polynomial Coefficients in Terms of the Green's Function

Tohru Morita [1],* and Ken-ichi Sato [2]

[1]   Graduate School of Information Sciences, Tohoku University, Sendai 980-8577, Japan
[2]   College of Engineering, Nihon University, Koriyama 963-8642, Japan; kensatokurume@ybb.ne.jp
*   Correspondence: senmm@jcom.home.ne.jp; Tel.: +81-22-278-6186

Received: 30 September 2017; Accepted: 6 November 2017; Published: 10 November 2017

Abstract: The particular solutions of inhomogeneous differential equations with polynomial coefficients in terms of the Green's function are obtained in the framework of distribution theory. In particular, discussions are given on Kummer's and the hypergeometric differential equation. Related discussions are given on the particular solution of differential equations with constant coefficients, by the Laplace transform.

Keywords: Green's function; distribution theory; particular solution; Kummer's differential equation; hypergeometric differential equation; Laplace transform

---

1. Introduction

In our recent papers [1,2], we are concerned with the solution of Kummer's and hypergeometric differential equations, where complementary solutions expressed by the confluent hypergeometric series and the hypergeometric series, respectively, are obtained, by using the Laplace transform, its analytic continuation, distribution theory and the fractional calculus.

It is the purpose of the present paper, to give the formulas which give the particular solutions of those equations with inhomogeneous term in terms of the Green's function. The differential equation satisfied by the Green's function is expressed with the aid of Dirac's delta function, which is defined in distribution theory, and hence the presentation in distribution theory is adopted.

Let

$$p_K(t, s) := t \cdot s^2 + (c - bt)s - ab, \tag{1}$$

where $a \in \mathbb{C}$, $b \in \mathbb{C}$ and $c \in \mathbb{C}$. Then Kummer's differential equation with an inhomogeneous term is given by

$$p_K(t, \frac{d}{dt})u(t) := t \cdot \frac{d^2}{dt^2}u(t) + (c - bt) \cdot \frac{d}{dt}u(t) - ab \cdot u(t) = f(t), \ \ t > 0. \tag{2}$$

If $c \notin \mathbb{Z}$, the basic complementary solutions of Equation (2) are given by

$$K_1(t) := {}_1F_1(a; c; bt), \tag{3}$$
$$K_2(t) := t^{1-c} \cdot {}_1F_1(a - c + 1; 2 - c; bt). \tag{4}$$

Here ${}_1F_1(a; c; z) = \sum_{k=0}^{\infty} \frac{(a)_k}{k!(c)_k} z^k$ is the confluent hypergeometric series, $(a)_k = \prod_{l=0}^{k-1}(a + l)$ for $k \in \mathbb{Z}_{>0}$, and $(a)_0 = 1$.

We use $\mathbb{R}$, $\mathbb{C}$ and $\mathbb{Z}$ to denote the sets of all real numbers, of all complex numbers and of all integers, respectively. We also use $\mathbb{R}_{>r} := \{x \in \mathbb{R} | x > r\}$, $\mathbb{R}_{\geq r} := \{x \in \mathbb{R} | x \geq r\}$ for $r \in \mathbb{R}$, $_+\mathbb{C} := \{z \in \mathbb{C} | \text{Re } z > 0\}$, $\mathbb{Z}_{>a} := \{n \in \mathbb{Z} | n > a\}$, $\mathbb{Z}_{<b} := \{n \in \mathbb{Z} | n < b\}$ and $\mathbb{Z}_{[a,b]} := \{n \in \mathbb{Z} | a \leq n \leq b\}$ for $a \in \mathbb{Z}$ and $b \in \mathbb{Z}$ satisfying $a < b$. We use Heaviside's step function $H(t)$, which is defined by

$$H(t) = \begin{cases} 1, & t > 0, \\ 0, & t \leq 0, \end{cases} \tag{5}$$

and when $f(t)$ is defined on $\mathbb{R}_{>\tau}$, $f(t)H(t-\tau)$ is equal to $f(t)$ for $t > \tau$ and to 0 for $t \leq \tau$.

When $c \in \mathbb{C}$ satisfies $\text{Re } (1-c) > -1$, the solution $K_2(t)$ has the Laplace transform:

$$\hat{K}_2(s) = \mathcal{L}[K_2(t)] := \int_0^\infty K_2(t)e^{-st}dt, \tag{6}$$

and is obtained by solving the Laplace transform of Equation (2). In [1,2], we confirm that the solution $K_2(t)$ is obtained by using an analytic continuation of the Laplace transform (AC-Laplace transform) for all nonzero values of $c \in \mathbb{C}\backslash\mathbb{Z}_{>0}$.

The complementary solution of the hypergeometric differential equation, corresponding to $K_2(t)$, is found to be obtained by using the Laplace transform series, where the Laplace transforms of the solutions are expresssed by a series of powers of s^{-1} multipied by a power of s, which has zero range of convergence. In fact, the series is the asymptotic expansion of Kummer's function $U(a, b, z)$ ([3], Section 13.5), which is discussed also in [4]. Even in that case, by the term-by-term inverse Laplace transform, we obtain the desired result. The calculation was justified by distribution theory [2].

In the present paper, we present the solutions giving particular solutions of Kummer's and the hypergeometric differential equation in terms of the Green's function with the aid of distribution theory.

In Section 2, we present the formulas in distribution theory, which are given in the book of Zemanian ([5], Section 6.3), where the particular solution of differential equation with constant coefficients is obtained. We use them in giving the particular solution of differential equation with polynomial coefficients in terms of the Green's function in Section 3, and the solutions are obtained by this method for Kummer's and the hypergeometric differential equation in Sections 4 and 5, respectively.

In Section 2.2, a formulation of the Laplace transform based on distribution theory is given, which is related with the one in ([5], Section 8.3).The particular solution of differential equation with constant coefficients is obtained by using the AC-Laplace transform in Section 6, which is compared with the formulation of the solution in distribution theory given in Section 3, where the Green's function plays an important role. In Sections 6.1 and 6.2, the solutions of differentaal equations of the first order and of the second order, respectively, are given.

We mention here that there are papers in which systematic study is made on polynomial solutions of inhomogeneous differential equation with polynomial coefficients; see [6] and its references. In the present paper, we are concerned with infinite series solutions. In our preceding papers [7,8] stimulated by Yosida's works [9,10] on Laplace's differential equations, of which typical one is Kummer's equation, we sudied the solution of Kummer's equation and a simple fractional differential equation on the basis of fractional calculus and distribution theory. In [1], we discussed it in terms of the AC-Laplace transform. In [4], we applied the arguments in [1] to the solution of the homogeneous hypergeometric equation. We now discuss the solution of inhomogeneous equations in terms of the Green's function and distribution theory. In [11,12], the solution of inhomogeneous differential equation with constant coefficients is discussed in terms of the Green's function and distribution theory. In Section 6, we discuss it in terms of the Green's function and the AC-Laplace transform, where we obtain the solution which is not obtained with the aid of the usual Laplace transform.

2. Preliminaries on Distribution Theory

Distributions in the space $\mathcal{D}'$ are first introduced in [5,13–15]. The distributions are either regular ones or their derivatives. A regular distribution $\tilde{f}$ in $\mathcal{D}'$ corresponds to a function f which belongs to $\mathcal{L}^1_{loc}(\mathbb{R})$. Here $\mathcal{L}^1_{loc}(\mathbb{R})$ denotes the class of functions which are locally integrable on $\mathbb{R}$. A distribution $\tilde{h}$, which is not a regular one, is expressed as $\tilde{h}(t) = D^n \tilde{f}(t)$, by $n \in \mathbb{Z}_{>0}$ and a regular distribution $\tilde{f} \in \mathcal{D}'$.

The space $\mathcal{D}$, that is dual to $\mathcal{D}'$, is the space of testing functions, which are infinitely differentiable on $\mathbb{R}$ and have a compact support.

Definition 1. *A distribution $\tilde{h} \in \mathcal{D}'$ is a functional, to which $\langle \tilde{h}, \phi \rangle \in \mathbb{C}$ is associated with every $\phi \in \mathcal{D}$. Let $\tilde{f}$ be the regular distribution which corresponds to a function $f \in \mathcal{L}^1_{loc}(\mathbb{R})$. Then (i):*

$$\langle \tilde{f}, \phi \rangle = \int_{-\infty}^{\infty} f(t)\phi(t)dt, \tag{7}$$

and (ii): if $n \in \mathbb{Z}_{>0}$, $\tilde{h}(t) = D^n \tilde{f}(t)$ is such a distribution belonging to $\mathcal{D}'$, that

$$\langle \tilde{h}, \phi \rangle = \langle D^n \tilde{f}, \phi \rangle = \langle \tilde{f}, D^n_W \phi \rangle = \int_{-\infty}^{\infty} f(t)[D^n_W \phi(t)]dt, \tag{8}$$

where $D^n_W \phi(t) = (-1)^n \frac{d^n}{dt^n}\phi(t)$.

Lemma 1. *Let $\tilde{f}$ be as in Definition 1, and $g(t) := \frac{d^n}{dt^n} f(t) = f^{(n)}(t) \in \mathcal{L}^1_{loc}(\mathbb{R})$, and the distribution which corresponds to $g(t) = f^{(n)}(t)$, be $\tilde{g}(t)$. Then $\tilde{g}(t) = D^n \tilde{f}(t)$.*

Proof. By using Equations (7) and (8), we have

$$\langle \tilde{g}, \phi \rangle = \int_{-\infty}^{\infty} [\frac{d^n}{dt^n} f(t)]\phi(t)dt = \int_{-\infty}^{\infty} f(t)[D^n_W \phi(t)]dt = \langle D^n \tilde{f}, \phi \rangle. \tag{9}$$

$\square$

Lemma 2. *D^n for $n \in \mathbb{Z}_{>-1}$ are operators in the space $\mathcal{D}'$.*

Definition 2. *Let $H(t)$ be defined by Equation (5), $\tau \in \mathbb{R}$, and $\psi(t)$ be such that $\psi(t)H(t-\tau) \in \mathcal{L}^1_{loc}(\mathbb{R})$. Then the regular distributions which correspond to $H(t-\tau)$ and $\psi(t)H(t-\tau)$ are denoted by $\tilde{H}(t-\tau)$ and $\psi(t)\tilde{H}(t-\tau)$, respectively.*

Lemma 3. *Let $\tau \in \mathbb{R}$, and $\psi(t)$ be such that $\psi(\tau) = 0$ and $\frac{d}{dt}[\psi(t)H(t-\tau)] = [\frac{d}{dt}\psi(t)]H(t-\tau) \in \mathcal{L}^1_{loc}(\mathbb{R})$. Then*

$$D[\psi(t)\tilde{H}(t-\tau)] = [\frac{d}{dt}\psi(t)]\tilde{H}(t-\tau). \tag{10}$$

Proof. We confirm this with the aid of Lemma 1 and Definition 2. $\square$

Dirac's delta function $\delta(t)$ is defined by

$$\delta(t) = D\tilde{H}(t). \tag{11}$$

Lemma 4. *Let $\tau \in \mathbb{R}$, and $\psi(t)$ be such that $\frac{d}{dt}\psi(t) \cdot H(t-\tau) \in \mathcal{L}^1_{loc}(\mathbb{R})$. Then*

$$D[\psi(t)\tilde{H}(t-\tau)] = \begin{cases} \frac{d}{dt}\psi(t) \cdot \tilde{H}(t-\tau), & \psi(\tau) = 0, \\ \frac{d}{dt}\psi(t) \cdot \tilde{H}(t-\tau) + \psi(\tau)\delta(t-\tau), & \psi(\tau) \neq 0. \end{cases} \tag{12}$$

Proof. The lhs is expressed as

$$D[\psi(t)\tilde{H}(t-\tau)] = D[[\psi(t) - \psi(\tau)]\tilde{H}(t-\tau)] + \psi(\tau)D\tilde{H}(t-\tau).$$

From this, we obtain Equation (12), with the aid of Lemma 3 and Equation (11). □

Corollary 1. *Let $\tau \in \mathbb{R}$, and $\psi(t)$ be such that $\frac{d^2}{dt^2}\psi(t) \cdot H(t-\tau) \in \mathcal{L}^1_{loc}(\mathbb{R})$. Then*

$$D^2[\psi(t)\tilde{H}(t-\tau)] = \frac{d^2}{dt^2}\psi(t) \cdot \tilde{H}(t-\tau) + \psi'(\tau)\delta(t-\tau) + \psi(\tau)D\delta(t-\tau). \tag{13}$$

Corollary 2. *Let $\tau \in \mathbb{R}$, $l \in \mathbb{Z}_{>0}$, $\psi(t)$ be such that $\frac{d^l}{dt^l}\psi(t) \cdot H(t-\tau) \in \mathcal{L}^1_{loc}(\mathbb{R})$, and $\tilde{\psi}_\tau(t) = \psi(t)\tilde{H}(t-\tau)$. Then*

$$D^l\tilde{\psi}_\tau(t) = D^l[\psi(t)\tilde{H}(t-\tau)] = \frac{d^l}{dt^l}\psi(t) \cdot \tilde{H}(t-\tau) + \langle D^l\hat{\psi}_\tau(D)\rangle_0\delta(t-\tau), \tag{14}$$

([5], Section 6.3), where

$$\langle D^l\hat{\psi}_\tau(D)\rangle_0 = \sum_{k=0}^{l-1} \psi^{(k)}(\tau)D^{l-k-1}. \tag{15}$$

When $l = 1$ and $l = 2$, we have

$$\langle D\hat{\psi}_\tau(D)\rangle_0 = \psi(\tau), \quad \langle D^2\hat{\psi}_\tau(D)\rangle_0 = \psi'(\tau) + \psi(\tau)D. \tag{16}$$

Equations (16) are obtained by comparing Equation (14) with Equations (12) and (13).

Corollary 3. *Let $\tau \in \mathbb{R}$, $n \in \mathbb{Z}_{>0}$, $\psi(t)$ be such that $\frac{d^n}{dt^n}\psi(t) \cdot H(t-\tau) \in \mathcal{L}^1_{loc}(\mathbb{R})$, $\tilde{\psi}_\tau(t) := \psi(t)\tilde{H}(t-\tau)$, and*

$$p_n(t, \frac{d}{dt})\psi(t) := \sum_{l=0}^n a_l(t)\frac{d^l}{dt^l}\psi(t), \tag{17}$$

where $a_l(t)$ are polynomials of t. Then

$$p_n(t, D)\tilde{\psi}_\tau(t) := \sum_{l=0}^n a_l(t)D^l\tilde{\psi}_\tau(t) = p_n(t, \frac{d}{dt})\psi(t) \cdot \tilde{H}(t-\tau) + \sum_{l=1}^n a_l(t)\langle D^l\hat{\psi}_\tau(D)\rangle_0\delta(t-\tau), \tag{18}$$

where $\langle D^l\hat{\psi}_\tau(D)\rangle_0$ are given by Equations (15) and (16).

In estimating the last term of Equation (18), we use the following lemma.

Lemma 5. *Let $\tilde{u}(t) \in \mathcal{D}'_R$, and $h(t)$ be such that $h(t)\phi(t) \in \mathcal{D}_R$ if $\phi \in \mathcal{D}_R$. Then*

$$h(t)D\tilde{u}(t) = D[h(t)\tilde{u}(t)] - h'(t)\tilde{u}(t). \tag{19}$$

Proof.

$$\langle h(t)D\tilde{u}(t), \phi(t)\rangle = \langle \tilde{u}(t), D_W[h(t)\phi(t)]\rangle = \langle -h'(t)\tilde{u}(t) + D[h(t)\tilde{u}(t)], \phi(t)\rangle, \tag{20}$$

since $D_W[h(t)\phi(t)] = -h'(t)\phi(t) + h(t)D_W\phi(t)$. □

2.1. Fractional Derivative and Distributions in the Space $\mathcal{D}'_R$

We consider the space $\mathcal{D}'_R$ [11,12]. A regular distribution in $\mathcal{D}'_R$ is such a distribution that it corresponds to a function which is locally integrable on $\mathbb{R}$ and has a support bounded on the left. A distribution $\tilde{h}$, which is not a regular one, is expressed as $\tilde{h}(t) = D^n \tilde{f}(t)$, by $n \in \mathbb{Z}_{>0}$ and a regular distribution $\tilde{f} \in \mathcal{D}'_R$. The space $\mathcal{D}_R$, that is dual to $\mathcal{D}'_R$, is the space of testing functions, which are infinitely differentiable on $\mathbb{R}$ and have a support bounded on the right.

We consider the Riemann-Liouville fractional integral and derivatives $_cD_R^\mu f(z)$ of order $\mu \in \mathbb{C}$, when we may usually discuss the derivative $\frac{d^n}{dz^n}f(z) = f^{(n)}(z)$ of order $n \in \mathbb{Z}_{>0}$; see [16] and ([17], Section 2.3.2). In the following definition, $P(c,z)$ is the path from $c \in \mathbb{C}$ to $z \in \mathbb{C}$, and $\mathcal{L}^1(P(c,z))$ is the class of functions which are integrable on $P(c,z)$, and $\lfloor x \rfloor$ for $x \in \mathbb{R}$ denotes the greatest integer not exceeding x. In the following study, we choose the value $c = 0$.

Definition 3. *Let $c \in \mathbb{C}$, $z \in \mathbb{C}$, $f(\zeta) \in \mathcal{L}^1(P(c,z))$, and $f(\zeta)$ be continuous in a neighborhood of $\zeta = z$. Then the Riemann-Liouville fractional integral of order $\lambda \in {}_+\mathbb{C}$ is defined by*

$$_cD_R^{-\lambda}f(z) = \frac{1}{\Gamma(\lambda)}\int_c^z (z-\zeta)^{\lambda-1}f(\zeta)\,d\zeta, \tag{21}$$

and the Riemann-Liouville fractional derivative of order $\mu \in \mathbb{C}$ satisfying Re $\mu \geq 0$ is defined by

$$_cD_R^\mu f(z) = {}_cD_R^m[_cD_R^{\mu-m}f(z)], \tag{22}$$

when the rhs exists, where $m = \lfloor \text{Re }\mu \rfloor + 1$, and $_cD_R^m f(z) = \frac{d^m}{dz^m}f(z) = f^{(m)}(z)$ for $m \in \mathbb{Z}_{>-1}$.

Definition 4. *Let $\lambda \in {}_+\mathbb{C}$, and $\tilde{f}$ be a regular distribution in $\mathcal{D}'_R$, that corresponds to a function f. Then we have a regular distribution which corresponds to $_0D_R^{-\lambda}f(t)$. We denote it $D^{-\lambda}\tilde{f}(t)$, and then*

$$\langle D^{-\lambda}\tilde{f}, \phi \rangle = \int_{-\infty}^\infty [_0D_R^{-\lambda}f(t)]\phi(t)dt = \langle \tilde{f}, D_W^{-\lambda}\phi \rangle = \int_{-\infty}^\infty f(x)[D_W^{-\lambda}\phi(x)]dx, \tag{23}$$

for every $\phi \in \mathcal{D}_R$. Then by using Equation (21) in the second member of Equation (23), we obtain

$$D_W^{-\lambda}\phi(x) = \frac{1}{\Gamma(\lambda)}\int_x^\infty (t-x)^{\lambda-1}\phi(t)\,dt. \tag{24}$$

This formula shows that $D_W^{-\lambda}\phi$ does not belong to $\mathcal{D}$, even when $\phi \in \mathcal{D}$. As a consequence, the operator $D^{-\lambda}$ corresponding $_cD_R^{-\lambda}$ cannot be an operator in the space $\mathcal{D}'$, but we can confirm that it is an operator in the space $\mathcal{D}'_R$.

Definition 5. *Let $\tilde{h}(t) \in \mathcal{D}'_R$, $m \in \mathbb{Z}_{>-1}$, $\lambda \in {}_+\mathbb{C}$, $\mu = m - \lambda$ and $D^\mu = D^m D^{-\lambda}$. Then we have $D^\mu\tilde{h} \in \mathcal{D}'_R$, which satisfies*

$$\langle D^\mu\tilde{h}, \phi \rangle = \langle \tilde{h}, D_W^\mu\phi \rangle, \tag{25}$$

for every $\phi \in \mathcal{D}_R$, where $D_W^\mu = D_W^{-\lambda}D_W^m$.

In solving a differential equation, we assume that the solution $u(t)$ and the inhomogeneous part $f(t)$ for $t > 0$, are expressed as a linear combination of

$$g_\nu(t) := \frac{1}{\Gamma(\nu)}t^{\nu-1}, \quad \nu \in \mathbb{C}\backslash\mathbb{Z}_{<1}, \tag{26}$$

where $\Gamma(\nu)$ is the gamma function. The Laplace transform of $g_\nu(t)$ is given by $\mathcal{L}[g_\nu(t)] = s^{-\nu}$ if $\nu \in {}_+\mathbb{C}$. We introduce the analytic continuation of the Laplace transform (AC-Laplace transform) of $g_\nu(t)$, which is expressed by $\mathcal{L}_H[g_\nu(t)]$, as in [1,2], such that

$$\hat{g}_\nu(s) = \mathcal{L}_H[g_\nu(t)] = s^{-\nu}, \quad \nu \in \mathbb{C}\backslash\mathbb{Z}_{<1}. \tag{27}$$

We often use the following formula [1,2].

Lemma 6. *Let $\nu \in \mathbb{C}\backslash\mathbb{Z}_{<1}$ and $\mu \in \mathbb{C}$. Then for $t \in \mathbb{R}_{>0}$, we have*

$$_0D_R^\mu g_\nu(t) = \begin{cases} g_{\nu-\mu}(t), & \nu - \mu \in \mathbb{C}\backslash\mathbb{Z}_{<1}, \\ 0, & \nu - \mu \in \mathbb{Z}_{<1}. \end{cases} \tag{28}$$

Condition 1. *A function of $t \in \mathbb{R}$ multiplied by $H(t)$, e.g. $u(t)H(t)$, is expressed as a linear combination of $g_\nu(t)H(t)$ for $\nu \in S$, where S is an enumerable set of $\nu \in \mathbb{C}\backslash\mathbb{Z}_{<1}$ satisfying $\text{Re } \nu > -M$ for some $M \in \mathbb{Z}_{>-1}$.*

When $u(t)$ satisfies Condition 1, it is expressed as follows:

$$u(t) = \sum_{\nu \in S} u_{\nu-1} g_\nu(t) = \sum_{\nu \in S} u_{\nu-1} \frac{1}{\Gamma(\nu)} t^{\nu-1}, \tag{29}$$

where $u_{\nu-1} \in \mathbb{C}$ are constants. When $\hat{u}(s) = \mathcal{L}_H[u(t)]$ exists, it is expressed by

$$\hat{u}(s) = \sum_{\nu \in S} u_{\nu-1} \hat{g}_\nu(s) = \sum_{\nu \in S} u_{\nu-1} s^{-\nu}. \tag{30}$$

Definition 6. *Let $g_\nu(t)$ and $\hat{g}_\nu(s)$ be given by Equations (26) and (27), respectively. Then $\tilde{g}_\nu(t) \in \mathcal{D}'_R$ is defined by*

$$\tilde{g}_\nu(t) = \hat{g}_\nu(D)\delta(t) = D^{-\nu}\delta(t), \quad \nu \in \mathbb{C}\backslash\mathbb{Z}_{<1}, \tag{31}$$

and is denoted by $g_\nu(t)\tilde{H}(t)$.

Remark 1. *When $\nu \in {}_+\mathbb{C}$, we can confirm that $\tilde{g}_\nu(t) = g_\nu(t)\tilde{H}(t)$ is a regular distribution which corresponds to $g_\nu(t)H(t)$, as*

$$\langle \tilde{g}_\nu, \phi \rangle = \langle D^{-\nu}\delta(t), \phi(t) \rangle = \langle D^{-\nu+1}\tilde{H}(t), \phi(t) \rangle = \int_{-\infty}^{\infty} H(t)[D_W^{-\nu+1}\phi(t)]dt$$

$$= \int_{-\infty}^{\infty} [_0D_R^{-\nu+1}H(t)]\phi(t)dt = \int_{-\infty}^{\infty} [g_\nu(t)H(t)]\phi(t)dt,$$

with the aid of Equations (9), (11) and (23), Definition 5, and formulas $g_1(t) = 1$, $g_\nu(t) = {}_0D_R^{-\nu+1}g_1(t)$ and $g_1(t) = {}_0D_R^{\nu-1}g_\nu(t)$ for $\nu \in \mathbb{C}\backslash\mathbb{Z}_{<1}$, which are obtained from Equations (26) and (28).

Lemma 7. *Let $g_\nu(t)$, $\hat{g}_\nu(s)$ and $\tilde{g}_\nu(t)$ be as in Definition 6. Then for $\mu \in \mathbb{C}$, we have*

$$D^\mu \tilde{g}_\nu(t) = {}_0D_R^\mu g_\nu(t) \cdot \tilde{H}(t) + \langle D^\mu \hat{g}_\nu(D) \rangle_0 \delta(t), \tag{32}$$

where

$$\langle D^\mu \hat{g}_\nu(D) \rangle_0 = \begin{cases} D^m, & m = \mu - \nu \in \mathbb{Z}_{>-1}, \\ 0, & \mu - \nu \notin \mathbb{Z}_{>-1}. \end{cases} \tag{33}$$

Proof. By using Equations (28) and (31), we confirm Equation (32). □

Lemma 8. *Let $u(t)$ and $\hat{u}(s) = \mathcal{L}_H[u(t)]$ be expressed by Equations (29) and (30), respectively. Then $\tilde{u}(t) = u(t)\tilde{H}(t) \in \mathcal{D}'_R$ is expressed as*

$$\tilde{u}(t) = \sum_{v \in S} u_{v-1}\tilde{g}_v(t) = \sum_{v \in S} u_{v-1}\hat{g}_v(D)\delta(t) = \hat{u}(D)\delta(t), \tag{34}$$

in accordance with Definition 6.

Lemma 9. *Let $l \in \mathbb{Z}_{>0}$, $u(t)$ be expressed by Equation (29), $\tilde{u}(t) := u(t)\tilde{H}(t)$, and $\hat{u}(s) := \mathcal{L}_H[u(t)]$. Then*

$$D^l\tilde{u}(t) = D^l[u(t)\tilde{H}(t)] = \frac{d^l}{dt^l}u(t) \cdot \tilde{H}(t) + \langle D^l\hat{u}(D)\rangle_0\delta(t), \tag{35}$$

where

$$\langle D^l\hat{u}(D)\rangle_0 = \sum_{k=0}^{l-1} u_k D^{l-k-1}. \tag{36}$$

When $l = 1, 2$ and 3, we have

$$\langle D\hat{u}(D)\rangle_0 = u_0, \quad \langle D^2\hat{u}(D)\rangle_0 = u_0 D + u_1, \quad \langle D^3\hat{u}(D)\rangle_0 = u_0 D^2 + u_1 D + u_2. \tag{37}$$

Proof. By using Equation (34) for $\tilde{u}(t)$, and Lemma 7 for $\mu = l \in \mathbb{Z}_{>0}$, we have contributions to $\langle D^l\hat{u}(D)\rangle_0\delta(t)$ from the terms of $v - 1 = k \in \mathbb{Z}_{>-1}$ and $m = l - k - 1 \in \mathbb{Z}_{>-1}$ in Equation (35). □

Remark 2. *Even when the series on the rhs of Equation (30) does not converge for any s, the series on the rhs of Equation (34) may converge in an interval of t on $\mathbb{R}$. In such a case, we use $\hat{u}(s)$ to represent the series on the rhs of Equation (30).*

2.2. Fractional Derivative and Distributions in the Space $\mathcal{D}'_W$

We adopt the space $\mathcal{D}'_W$ of distributions, such that regular ones correspond to functions which may increase slower than $e^{\epsilon t}$ for all $\epsilon \in \mathbb{R}_{>0}$ as $t \to \infty$. Then the dual space $\mathcal{D}_W$ consists of functions which are infinitely differentiable on $\mathbb{R}$ and decay more rapidly than t^{-N} for all $N \in \mathbb{Z}_{>0}$ as $t \to \infty$.

Remark 3. *In the book of Miller and Ross ([18], Chapter VII), one chapter is used to discuss the Weyl fractional integral and differentiation in the space $\mathcal{D}_W$, where notation $W^\mu g(t)$ is used in place of $D^\mu_W g(t)$ for $\mu \in \mathbb{R}$ and $\phi \in \mathcal{D}_W$.*

Lemma 10. *Let $m \in \mathbb{Z}_{>-1}$, $\lambda \in {}_+\mathbb{C}$ and $\mu = m - \lambda$. Then*

$$D^m_W e^{-st} = s^m e^{-st}, \quad D^{-\lambda}_W e^{-st} = s^{-\lambda} e^{-st}, \quad D^\mu_W e^{-st} = s^\mu e^{-st}. \tag{38}$$

Proof. The second equation is confirmed with the aid of Equation (24). The third equation is confirmed with the aid of Definition 5, which states $D^\mu_W = D^{-\lambda}_W D^m_W$. □

Lemma 11. *Let $\tilde{u}(t) = u(t)\tilde{H}(t)$ be a regular distribution in $\mathcal{D}'_W$, that corresponds to function $u(t)H(t)$, $\hat{u}(s) = \mathcal{L}[u(t)] = \int_0^\infty u(t)e^{-st}dt$, and $\tilde{w}(t)$ and $\hat{w}(s)$ be given by $\tilde{w}(t) = D^l\tilde{u}(t)$ and $\hat{w}(s) = s^l\hat{u}(s)$ for $l \in \mathbb{Z}_{>0}$. Then for $\mu \in \mathbb{C}$, we have*

$$\text{(i) } \langle \tilde{u}(t), e^{-st}\rangle = \hat{u}(s), \quad \text{(ii) } \langle D^\mu\tilde{u}(t), e^{-st}\rangle = s^\mu\hat{u}(s), \quad \text{(iii) } \langle tD^\mu\tilde{u}(t), e^{-st}\rangle = -\frac{d}{ds}[s^\mu\hat{u}(s)], \tag{39}$$

and also these equations with $\tilde{u}(t)$ and $\hat{u}(s)$ replaced by $\tilde{w}(t)$ and $\hat{w}(s)$, respectively.

Proof. The lhs of these equations are expressed as follows: (i) $\langle \tilde{u}(t), e^{-st} \rangle = \int_{-\infty}^{\infty} u(t)H(t)e^{-st}dt = \mathcal{L}[u(t)] = \hat{u}(s)$, (ii) $\langle D^\mu \tilde{u}(t), e^{-st} \rangle = \langle \tilde{u}(t), D_W^\mu e^{-st} \rangle = s^\mu \langle \tilde{u}(t), e^{-st} \rangle = s^\mu \hat{u}(s)$, and (iii) $\langle tD^\mu \tilde{u}(t), e^{-st} \rangle = \langle D^\mu \tilde{u}(t), te^{-st} \rangle = -\frac{d}{ds}\langle D^\mu \tilde{u}(t), e^{-st} \rangle = -\frac{d}{ds}[s^\mu \hat{u}(s)]$. $\square$

This lemma shows that the formulas in Equation (39) are valid for all distributions $\tilde{u}(t) \in \mathcal{D}'_W$.

Corollary 4. *Let the condition of Lemma 6 be satisfied. Then* $\delta(t) \in \mathcal{D}'_W$, $\tilde{g}_\nu(t) \in \mathcal{D}'_W$ *for* $\nu \in \mathbb{C} \backslash \mathbb{Z}_{<1}$, *and*

$$\langle \tilde{H}(t), e^{-st} \rangle = \frac{1}{s}; \langle \delta(t), e^{-st} \rangle = 1;$$
$$\langle \tilde{g}_\nu(t), e^{-st} \rangle = \langle \hat{g}_\nu(D)\delta(t), e^{-st} \rangle = \hat{g}_\nu(s) = s^{-\nu}. \tag{40}$$

Proof. $\langle \delta(t), e^{-st} \rangle = \langle D\tilde{H}(t), e^{-st} \rangle = \langle \tilde{H}(t), D_W e^{-st} \rangle = s^{-1}s = 1$. $\square$

Corollary 5. *Let* $\tilde{u}(t) \in \mathcal{D}'_W$ *be expressed as* $\tilde{u}(t) = \hat{u}(D)\delta(t)$. *Then*

$$\langle \tilde{u}(t), e^{-st} \rangle = \langle \hat{u}(D)\delta(t), e^{-st} \rangle = \hat{u}(s). \tag{41}$$

Lemma 12. *Let* $l \in \mathbb{Z}_{>0}$, $u(t)$ *be expressed by Equation* (29), *and* $\hat{u}(s) := \mathcal{L}_H[u(t)]$. *Then*

$$s^l \hat{u}(s) = \mathcal{L}_H[\frac{d^l}{dt^l}u(t)] + \langle s^l \hat{u}(s) \rangle_0, \tag{42}$$

where $\langle s^l \hat{u}(s) \rangle_0$ *are given by Equations* (36) *and* (37) *with D replaced by s.*

Proof. We put $\tilde{u}(t) = u(t)\tilde{H}(t)$, and then we have Equation (35). Applying Lemma 11 and Corollary 5 to Equation (35), we obtain Equation (42). $\square$

Remark 4. *In the book by Zemanian ([5], Section 8.3), he discussed the Laplace transform by adopting the space* $\mathcal{D}'_L$ *of distributions, such that regular ones correspond to functions which may increase slower than* $e^{(\lambda+\epsilon)t}$ *for all* $\epsilon \in \mathbb{R}_{>0}$ *as* $t \to \infty$, *for a fixed* $\lambda \in \mathbb{R}_{>0}$. *Then the dual space* $\mathcal{D}_L$ *consists of functions which are infinitely differentiable on* $\mathbb{R}$ *and decay more rapidly than* $e^{-\lambda t}t^{-N}$ *for all* $N \in \mathbb{Z}_{>0}$ *as* $t \to \infty$. *Then the above formulas in the present section are valid when* Re $s > \lambda$. *We here adopt the case when* $\lambda = 0$.

Remark 5. *Let* $u(t)$ *and* $\tilde{u}(t)$ *be expressed by Equations* (29) *and* (34), *respectively. Then* $\tilde{u}(t)$ *does not usually belongs to* $\mathcal{D}'_W$. *Even in that case, we use the following equation:*

$$\langle \tilde{u}(t), e^{-st} \rangle = \langle \hat{u}(D)\delta(t), e^{-st} \rangle = \sum_{\nu \in S} u_{\nu-1}\langle \tilde{g}_\nu(t), e^{-st} \rangle = \sum_{\nu \in S} u_{\nu-1}\hat{g}_\nu(s) = \hat{u}(s). \tag{43}$$

2.3. Some Primitive Leibniz's formulas

Lemma 13. *Let* $\tilde{u}(t) = \hat{u}(D)\delta(t) \in \mathcal{D}'_R$, *and* $\mu \in \mathbb{C}$. *Then we have two special ones of Leibniz's formula:*

$$D^\mu[t\tilde{u}(t)] = tD^\mu \tilde{u}(t) + \mu D^{\mu-1}\tilde{u}(t) = (tD + \mu)D^{\mu-1}\tilde{u}(t), \tag{44}$$

$$D^\mu[t^2\tilde{u}(t)] = t^2 D^\mu \tilde{u}(t) + 2\mu t D^{\mu-1}\tilde{u}(t) + \mu(\mu-1)D^{\mu-2}\tilde{u}(t)$$
$$= (t^2 D^2 + 2\mu t D + \mu(\mu-1))D^{\mu-2}\tilde{u}(t). \tag{45}$$

Proof. Lemma 11 shows that when Equation (44) is satisfied, we have

$$s^\mu[-\frac{d}{ds}\hat{u}(s)] = -\frac{d}{ds}[s^\mu \hat{u}(s)] + \mu s^{\mu-1}\hat{u}(s), \tag{46}$$

which is confirmed since $-\frac{d}{ds}[s^\mu \hat{u}(s)] = -\mu s^{\mu-1}\hat{u}(s) - s^\mu \frac{d}{ds}\hat{u}(s)$. Formula (45) is obtained with the aid of Equation (44). $\square$

3. Green's Function for Inhomogeneous Differential Equations with Polynomial Coefficients

In Sections 4 and 5, we study the solution $u(t)$ of inhomogeneous differential equations with polynomial coefficients:

$$p_n(t, \frac{d}{dt})u(t) := \sum_{l=0}^{n} a_l(t)\frac{d^l}{dt^l}u(t) = f(t), \quad t > 0, \tag{47}$$

where $n \in \mathbb{Z}_{>0}$, and $a_l(t)$ are polynomials of t satisfying $a_0(t) \neq 0$ and $a_n(t) \neq 0$.

For the inhomogeneous term $f(t)$, we consider the following three cases.

Condition 2.

(i) $f(t)$ is a locally integrable function on $\mathbb{R}_{\geq 0}$,

(ii) $f(t) = {}_0 D_R^\beta f_\beta(t)$ satisfies Condition 1, where $f_\beta(t)$ is a locally integrable function on $\mathbb{R}_{\geq 0}$,

(iii) $f(t) = g_\nu(t) = \frac{1}{\Gamma(\nu)}t^{\nu-1}$, and $\tilde{f}(t) = f(t)\tilde{H}(t) = \hat{f}(D)\delta(t) = \hat{g}_\nu(D)\delta(t) = D^{-\nu}\delta(t)$, for $\nu \in \mathbb{C}\backslash\mathbb{Z}_{<1}$.

Lemma 14. *Let $u(t)$ be a solution of Equation (47), which is expressed by Equation (29). Then the differential equation satisfied by distribution $\tilde{u}(t) := u(t)\tilde{H}(t)$ is*

$$\begin{aligned}
p_n(t, D)\tilde{u}(t) &:= \sum_{l=0}^{n} a_l(t)D^l\tilde{u}(t) = p_n(t, \tfrac{d}{dt})u(t) \cdot \tilde{H}(t) + \sum_{l=1}^{n} a_l(t)\langle D^l\hat{u}(D)\rangle_0 \delta(t) \\
&= \tilde{f}(t) + \sum_{l=1}^{n} a_l(t)\langle D^l\hat{u}(D)\rangle \delta(t),
\end{aligned} \tag{48}$$

where $\tilde{f}(t) = f(t)\tilde{H}(t)$, and $\langle D^l\hat{u}(D)\rangle_0$ are given by Equations (36) and (37).

Proof. Equation (48) is obtained by using Equation (35). $\square$

Lemma 15. *Let Condition 2(i) be satisfied. Then $\tilde{f}(t) = f(t)\tilde{H}(t)$, which corresponds to $f(t)H(t)$, is expressed as*

$$\tilde{f}(t) = \int_0^\infty \delta(t - \tau)f(\tau)d\tau. \tag{49}$$

Proof. The rhs is equal to the lhs, since

$$\begin{aligned}
\left\langle \int_0^\infty \delta(t-\tau)f(\tau)d\tau, \phi(t) \right\rangle &= -\int_{-\infty}^{\infty}\left[\int_0^\infty H(t-\tau)f(\tau)d\tau\right]\phi'(t)dt \\
&= -\int_0^\infty f(\tau)\left[\int_{-\infty}^{\infty} H(t-\tau)\phi'(t)dt\right]d\tau = \int_{-\infty}^{\infty} f(\tau)H(\tau)\phi(\tau)d\tau = \langle\tilde{f},\phi\rangle,
\end{aligned}$$

where $\phi \in \mathcal{D}_R$. $\square$

Definition 7. *For Equation (47), the Green's function $G(t,\tau)$ for $\tau \in \mathbb{R}_{\geq 0}$ is such that $G(t,\tau) = G(t,\tau)H(t-\tau)$, and $\tilde{G}(t,\tau) = G(t,\tau)\tilde{H}(t-\tau)$ satisfies*

$$p_n(t, D)\tilde{G}(t,\tau) = \delta(t - \tau). \tag{50}$$

Lemma 16. *The Green's function $G(t,\tau)$ for Equation (47) satisfies*

$$p_n(t, \frac{d}{dt})G(t,\tau) = \sum_{l=0}^{n} a_l(t)\frac{d^l}{dt^l}G(t,\tau) = 0, \quad t > \tau, \tag{51}$$

and $G(t, \tau) = 0$ for $t < \tau$. The values of $G(t, \tau)$ and its derivatives at $t = \tau$ are determined by

$$\sum_{l=1}^{n} a_l(t) \langle D^l \hat{G}(D, \tau) \rangle_0 \delta(t - \tau) = \delta(t - \tau), \tag{52}$$

where $\langle D^l \hat{G}(D, \tau) \rangle_0$ is given by Equations (15) and (16) with $\hat{\psi}_\tau(D)$ and $\tilde{\psi}_\tau(t)$ replaced by $\hat{G}(D, \tau)$ and $\tilde{G}(t, \tau)$, respectively.

Proof. This is confirmed by comparing Equation (50) with Equation (18), where $\psi(t)$, $\hat{\psi}_\tau(D)$ and $\tilde{\psi}_\tau(t)$ are replaced by $G(t, \tau)$, $\hat{G}(D, \tau)$ and $\tilde{G}(t, \tau)$, respectively. $\square$

Lemma 17. *$\tilde{u}_f(t)$ given by*

$$\tilde{u}_f(t) = \int_0^\infty \tilde{G}(t, \tau) f(\tau) d\tau \tag{53}$$

is a particular solution of Equation (48) for the term $\tilde{f}(t)$.

Proof. By using the rhs of Equation (53) in the lhs of Equation (48), we obtain $f(t)\tilde{H}(t)$ by using Equations (50) and (49). $\square$

Lemma 18. *Let $G(t, \tau)$ be defined by Definition 7 for Equation (47), Condition 2(i) be satisfied, and $u_f(t)$ be given by*

$$u_f(t) = \int_0^t G(t, \tau) f(\tau) d\tau. \tag{54}$$

Then $u_f(t)$ and $\tilde{u}_f(t) = u_f(t)\tilde{H}(t)$ are particular solutions of Equations (47) and (48), respectively.

Proof. We show that when $\tilde{u}_f(t)$ and $u_f(t)$ are defined by Equations (53) and (54), respectively, $\tilde{u}_f(t) = u_f(t)\tilde{H}(t)$. This is confirmed as follows:

$$\begin{aligned}
\langle \tilde{u}_f(t), \phi \rangle &= \int_0^\infty \left[\int_{-\infty}^\infty G(t, \tau) H(t - \tau) \phi(t) dt \right] f(\tau) d\tau \\
&= \int_{-\infty}^\infty \left[\int_0^t G(t, \tau) f(\tau) d\tau \right] H(t) \phi(t) dt = \langle u_f(t)\tilde{H}(t), \phi(t) \rangle,
\end{aligned} \tag{55}$$

where $\phi \in \mathcal{D}_R$. The proof of the lemma is completed with the aid of Lemmas 17 and 14. $\square$

Green's Function for Inhomogeneous Differential Equations with Constant Coefficients

We now consider the case when $a_l(t)$ do not depend on t, and hence in place of Equation (47), we have

$$p_n(\frac{d}{dt})u(t) := \sum_{l=0}^{n} a_l \frac{d^l}{dt^l} u(t) = f(t), \quad t > 0, \tag{56}$$

where $n \in \mathbb{Z}_{>0}$, and $a_l \in \mathbb{C}$ are constants satisfying $a_0 \neq 0$ and $a_n \neq 0$. When we use the formulas in the preceding section, we assume that $a_l(t)$ and $p_n(t, \frac{d}{dt})$ represent a_l and $p_n(\frac{d}{dt})$, respectively. In place of Equation (48), we have

$$p_n(D)\tilde{u}(t) := \sum_{l=0}^{n} a_l D^l \tilde{u}(t) = \tilde{f}(t) + \sum_{l=1}^{n} a_l \langle D^l \hat{u}(D) \rangle_0 \delta(t), \tag{57}$$

where $\tilde{f}(t) = f(t)\tilde{H}(t) = \hat{f}(D)\delta(t)$, and $\langle D^l \hat{u}(D) \rangle_0$ are given by Equations (36) and (37).

We denote the solution $G(t, 0)$ obtained by Definition 7, by $G(t)$. Then $\tilde{G}(t) = G(t)\tilde{H}(t)$ satisfies

$$p_n(D)\tilde{G}(t) = \delta(t). \tag{58}$$

By using this equation, we confirm the following lemma.

Lemma 19. *$\tilde{u}(t)$ given by*

$$\tilde{u}(t) = \hat{f}(D)\tilde{G}(t) + \sum_{l=1}^{n} a_l \langle D^l \hat{u}(D) \rangle_0 \tilde{G}(t), \tag{59}$$

is a solution of Equation (57).

Proof. Putting $\tilde{u}(t)$ given by Equation (59) on the lhs of Equation (57) and using Equation (58), we confirm that Equation (57) is satisfied. □

By Lemma 16, we have

Lemma 20. *Let $G(t)$ be defined by*

$$p_n(\frac{d}{dt})G(t) := \sum_{l=0}^{n} a_l \frac{d^l}{dt^l} G(t) = \frac{d}{dt}H(t) = 0, \quad t > 0, \tag{60}$$

and the initial values of $G(t)$ and its derivatives satisfy $\sum_{l=1}^{n} a_l \langle D^l \hat{G}(D) \rangle_0 = 1$, so that

$$a_n G^{(n-1)}(0) = H(0) = 1; \quad G^{(l)}(0) = 0, \quad (l \in \mathbb{Z}_{[0,n-2]}, \ n \in \mathbb{Z}_{>1}). \tag{61}$$

Then $\tilde{G}(t) = G(t)\tilde{H}(t)$ satisfies (58), and $\tilde{G}(t, \tau) = G(t - \tau)H(t - \tau)$ for $\tau \geq 0$ satisfies Definition 7 if $p_n(t, D) = p_n(D)$.

Proof. By comparing Equations (50) and (58), we confirm the last statement. □

Lemma 21. *Let Condition 2(i) be satisfied, and $u(t)$ be given by*

$$u(t) = u_f(t) + \sum_{l=1}^{n} a_l \langle \frac{d^l}{dt^l} \hat{u}(\frac{d}{dt}) \rangle_0 G(t), \tag{62}$$

where

$$u_f(t) = \int_0^t G(t - \tau)f(\tau)d\tau. \tag{63}$$

Then $u(t)$ and $\tilde{u}(t) = u_f(t)\tilde{H}(t)$ are solutions of Equations (56) and (57), respectively.

Proof. Equation (63) follows from Lemma 18. □

Remark 6. *In ([5], Section 6.3), the solution of an inhomogeneous differential equation with constant coefficients is discussed, where the formulas in Corollary 2, Formulas (60) and (61) for the Green's function and Formula (62) for the solution, are presented.*

4. Particular Solution of Kummer's Differential Equation

Kummer's differential equation with an inhomogeneous term $f(t)$ is given in Equation (2). If $f(t) = 0$ and $c \notin \mathbb{Z}$, the basic solutions $K_1(t)$ and $K_2(t)$ of Equation (2) are given by Equations (3) and (4).

We now obtain a particular solution of this equation by the method stated in Section 3.

4.1. Green's Function for Kummer's Differential Equation in which Condition 2(i) Is Satisfied

The following two lemmas are proved in Section 4.4.

Lemma 22. *Let $u(t)$ be the solution of Equation (2), which is expressed by Equation (29). Then the differential equation satisfied by $\tilde{u}(t) = u(t)\tilde{H}(t)$ is given by*

$$p_K(t, D)\tilde{u}(t) = f(t)\tilde{H}(t) + u_0(c-1)\delta(t). \tag{64}$$

Lemma 23. *Let $K_1(t)$ and $K_2(t)$ be given by Equations (3) and (4), and $\psi_K(t, \tau)$ for fixed $\tau > 0$ be given by*

$$\psi_K(t, \tau) := c_1(\tau) \cdot K_1(t) + c_2(\tau) \cdot K_2(t), \tag{65}$$

where $c_1(\tau)$ and $c_2(\tau)$ are constants which are so chosen that $\psi_K(t, \tau) = 0$ when $t = \tau$. Then $G_K(t, 0)$ and $G_K(t, \tau)$ given by

$$G_K(t, 0) = \frac{1}{c-1} \cdot K_1(t)H(t), \quad G_K(t, \tau) = \frac{1}{\tau\psi'_K(\tau, \tau)}\psi_K(t, \tau)H(t-\tau), \tag{66}$$

are the Green's functions for Equation (2), so that $\tilde{G}_K(t, 0) = G_K(t, 0)\tilde{H}(t)$ and $\tilde{G}_K(t, \tau) = G_K(t, \tau)\tilde{H}(t-\tau)$ satisfy

$$p_K(t, D)\tilde{G}_K(t, 0) = \delta(t), \quad p_K(t, D)\tilde{G}_K(t, \tau) = \delta(t-\tau). \tag{67}$$

Theorem 1. *Let $G_K(t, \tau)$ and $\psi_K(t, \tau)$ be those given in Lemma 23, Condition 2(i) be satisfied, and $u_f(t)$ be given by*

$$u_f(t) = \int_0^t G_K(t, \tau)f(\tau)d\tau = \int_0^t \psi_K(t, \tau)\frac{f(\tau)}{\tau\psi'_K(\tau, \tau)}d\tau. \tag{68}$$

Then $u_f(t)$ and $u_f(t)\tilde{H}(t)$ are particular solutions of Equations (2) and (64) for the terms $f(t)$ and $f(t)\tilde{H}(t)$, respectively.

Proof. This is confirmed with the aid of Lemma 18. □

Remark 7. *By using the first equation in Equation (67), we see that the particular solution of Equation (64) for the last term, is*

$$\tilde{u}_1(t) = u_0(c-1)\tilde{G}_K(t, 0) = u_0 \cdot K_1(t)\tilde{H}(t). \tag{69}$$

The corresponding particular solution of Equation (2) is

$$u_1(t) = u_0(c-1)G_K(t, 0) = u_0 \cdot K_1(t). \tag{70}$$

Considering that the basic complementary solutions of Equation (2) are given by Equations (3) and (4), the general solution of Equation (2) is now given by

$$u(t) = u_f(t) + u_0 \cdot K_1(t) + c_3 \cdot K_2(t). \tag{71}$$

The condition $\psi_K(\tau, \tau) = 0$ requires that $c_1(\tau) \cdot K_1(\tau) = -c_2(\tau) \cdot K_2(\tau)$, and hence we may choose $\psi_K(t, \tau)$ as

$$\psi_K(t, \tau) = \begin{cases} K_1(t) - \frac{K_1(\tau)}{K_2(\tau)} \cdot K_2(t), & |K_1(\tau)| < |K_2(\tau)|, \\ \frac{K_2(\tau)}{K_1(\tau)} \cdot K_1(t) - K_2(t), & |K_1(\tau)| \geq |K_2(\tau)|. \end{cases} \tag{72}$$

4.2. Green's Function for Kummer's Differential Equation in which Condition 2(ii) Is Satisfied

We give the solution of Equation (2) of which the inhomogeneous term $f(t)$ satisfies Condition 2(ii), so that $\tilde{f}(t) = D^\beta \tilde{f}_\beta(t)$, and $f_\beta(t)$ satisfies Condition 2(i).

The following lemma is proved in Section 4.4.

Lemma 24. *Let $\tilde{u}(t)$ be a solution of Equation (64), and $p_{\tilde{R}}(t, D)$ be related with $p_K(t, D)$ given by Equation (1), by*

$$p_{\tilde{R}}(t, D) = D^{-\beta} p_K(t, D) D^\beta. \tag{73}$$

Then

$$p_{\tilde{R}}(t, D) := t \cdot D^2 + (c - \beta - bt)D - (a - \beta)b, \tag{74}$$

and $\tilde{w}(t) = D^{-\beta}\tilde{u}(t)$ satisfies

$$p_{\tilde{R}}(t, D)\tilde{w}(t) = f_\beta(t)\tilde{H}(t) + u_0(c - 1)D^{-\beta}\delta(t). \tag{75}$$

Theorem 1 shows that the particular solution of Equation (75) for the term $f_\beta(t)\tilde{H}(t)$ is expressed by a particular solution of

$$p_{\tilde{R}}(t, \frac{d}{dt})w(t) = f_\beta(t), \quad t > 0. \tag{76}$$

That solution is used in giving a solution of Equation (2) in Theorem 2 given below.

Remark 8. *We note that $p_{\tilde{R}}(t, D)$ given by Equation (74) is obtained from $p_K(t, D)$ given by Equation (1), by replacing a and c by $a - \beta$ and $c - \beta$, respectively, and hence the complementary solutions $K_{\beta,1}(t)$ and $K_{\beta,2}(t)$ and Green's functions $G_{\tilde{R}}(t, 0)$ and $G_{\tilde{R}}(t, \tau)$ of Equation (76) are obtained from those $K_1(t)$, $K_2(t)$, $G_K(t, 0)$ and $G_K(t, \tau)$, respectively, of Equation (2), by the same replacement.*

Now in place of Theorem 1, we have the following theorem, whose proof is given in Section 4.4.

Theorem 2. *Let $G_{\tilde{R}}(t, \tau)$ and $\psi_{\tilde{R}}(t, \tau)$ be obtained from $G_K(t, \tau)$ and $\psi_K(t, \tau)$ by the replacement stated in Remark 8, Condition 2(ii) be satisfied, and $w_g(t)$ be given by*

$$w_g(t) = \int_0^t G_{\tilde{R}}(t, \tau)f_\beta(\tau)d\tau = \int_0^t \psi_{\tilde{R}}(t, \tau)\frac{f_\beta(\tau)}{\tau\psi'_{\tilde{R}}(\tau, \tau)}d\tau. \tag{77}$$

Then $u_f(t) := {}_0D_R^\beta w_g(t)$ and $u_f(t)\tilde{H}(t)$ are particular solutions of Equations (2) and (64) for the terms $f(t)$ and $f(t)\tilde{H}(t)$, respectively.

4.3. Particular Solution of Kummer's Differential Equation in which Condition 2(iii) Is Satisfied

We give the solution of Equation (2) of which the inhomogeneous term $f(t)$ satisfies Condition 2(iii), so that $f(t) = g_\nu(t) = \frac{1}{\Gamma(\nu)}t^{\nu-1}$ and $\tilde{f}(t) = f(t)\tilde{H}(t) = \hat{f}(D)\delta(t) = \hat{g}_\nu(D)\delta(t) = D^{-\nu}\delta(t)$. Here we use $\beta \notin \mathbb{Z}_{>-1}$ in place of $-\nu$.

Lemma 25. *Let* $\tilde{f}(t) = D^{\beta}\delta(t)$, $p_{\tilde{R}}(t, D)$ *be given by Equation* (74), *and* $\tilde{G}_{\tilde{R}}(t, 0)$ *satisfy*

$$p_{\tilde{R}}(t, D)\tilde{G}_{\tilde{R}}(t, 0) = \delta(t). \tag{78}$$

Then the particular solution of Equation (64) *for the term* $f(t)\tilde{H}(t) = D^{\beta}\delta(t)$ *is given by*

$$\tilde{u}_f(t) = D^{\beta}\tilde{G}_{\tilde{R}}(t, 0). \tag{79}$$

Proof. $\tilde{u}_f(t)$ satisfies $p_K(t, D)\tilde{u}_f(t) = D^{\beta}\delta(t)$ and hence $p_{\tilde{R}}(t, D)D^{-\beta}\tilde{u}_f(t) = \delta(t)$ by Equation (73). Comparing this with Equation (78), we see that Equation (79) is satisfied. □

By Remark 8, with the aid of $G_K(t, 0)$ given by Equations (66) and (3), we have

Lemma 26. *As the solution of Equation* (78), *we obtain*

$$\tilde{G}_{\tilde{R}}(t, 0) = G_{\tilde{R}}(t, 0)\tilde{H}(t), \quad G_{\tilde{R}}(t, 0) = \frac{1}{c - \beta - 1} \cdot {}_1F_1(a - \beta; c - \beta; bt). \tag{80}$$

Theorem 3. *Let Condition* 2*(iii) be satisfied, and* $G_{\tilde{R}}(t, 0)$ *be given by Equation* (80). *Then the particular solution of Equation* (2) *is given by*

$$u_f(t) = {}_0D^{\beta}_R G_{\tilde{R}}(t, 0) = \frac{t^{-\beta}}{(c - \beta - 1)\Gamma(1 - \beta)} \cdot {}_2F_2(1, a - \beta; 1 - \beta, c - \beta; bt), \tag{81}$$

where ${}_2F_2(a_1, a_2; c_1, c_2; z) = \sum_{k=0}^{\infty} \frac{(a_1)_k(a_2)_k}{k!(c_1)_k(c_2)_k} z^k$.

Proof. $\tilde{u}_f(t)$ is given by Equations (79) and (80). Lemma 7 shows that $\tilde{u}_f(t) = u_f(t)\tilde{H}(t)$ if $u_f(t)$ is given by Equation (81), since $\beta \notin \mathbb{Z}_{>-1}$. □

4.4. Proofs of Lemmas 22–24 *and Theorem* 2

Proof of Lemma 22. Equation (64) is a special one of (48), where p_n, n, $a_2(t)$, $a_1(t)$ and $a_0(t)$ stand for p_K, 2, t, $c - bt$ and $-ab$, respectively. Hence by using Equation (37), we see that the second term on the rhs of Equation (48) becomes

$$t(u_0 D + u_1)\delta(t) + (c - bt)u_0\delta(t) = u_0(D[t\delta(t)] - \delta(t)) + cu_0\delta(t) = u_0(c - 1)\delta(t). \tag{82}$$

□

We prepare a lemma before proving Lemma 23.

Lemma 27. *Let* $\tau \geq 0$, $p_K(t, \frac{d}{dt})\psi(t) \cdot H(t - \tau) \in \mathcal{L}^1_{loc}$ *and* $\tilde{\psi}_{\tau}(t) = \psi(t)\tilde{H}(t - \tau)$. *Then*

$$p_K(t, D)\tilde{\psi}_{\tau}(t) = p_K(t, \frac{d}{dt})\psi(t) \cdot \tilde{H}(t - \tau) + \sum_{l=1}^{2} a_l(t)\langle D^l\hat{\psi}_{\tau}(D)\rangle_0\delta(t - \tau), \tag{83}$$

where $a_2(t) = t$, $a_1(t) = c - bt$, $a_0(t) = ab$, *and*

$$\sum_{l=1}^{2} a_l(t)\langle D^l\hat{\psi}_{\tau}(D)\rangle_0\delta(t - \tau) = \tau\psi'(\tau)\delta(t - \tau) + \psi(\tau)(c - b\tau - 1)\delta(t - \tau) + \psi(\tau)\tau D\delta(t - \tau). \tag{84}$$

Proof. Equation (83) is a special one of Equation (18) in Corollary 3. The lhs of Equation (84) is evaluated with the aid of Equation (16) as follows:

$$\sum_{l=1}^{2} a_l(t)\langle D^l \hat{\psi}_\tau(D)\rangle_0 \delta(t-\tau) = t[\psi'(\tau) + \psi(\tau)D]\delta(t-\tau) + (c - bt)\psi(\tau)\delta(t-\tau)$$

$$= [t\psi'(\tau) + \psi(\tau)(c-bt)]\delta(t-\tau) + \psi(\tau)tD\delta(t-\tau)$$

$$= \tau\psi'(\tau)\delta(t-\tau) + \psi(\tau)(c - b\tau - 1)\delta(t-\tau) + \psi(\tau)\tau D\delta(t-\tau),$$

where Formula (19) with $h(t)$ and $\tilde{u}(t)$ replaced by t and $\delta(t-\tau)$, respectively, is used in the last term of the third member. $\square$

Proof of Lemma 23. If we put $\psi(t) = \frac{1}{c-1} \cdot K_1(t)$ and $\tau = 0$ in Equation (83), then we have $\hat{\psi}_\tau(t) = \tilde{G}(t,0)$ on the lhs, and the rhs is $\psi(0)(c-1)\delta(t) = \delta(t)$, since $K_1(0) = 1$. If we put $\psi(t) = \frac{1}{\tau\psi'_K(\tau,\tau)}\psi_K(t,\tau)$ and $\psi(\tau) = 0$ in Equation (83), then we have $\hat{\psi}_\tau(t) = \tilde{G}(t,\tau)$ on the lhs, and the rhs is $\tau\psi'(\tau)\delta(t-\tau) = \delta(t-\tau)$. $\square$

Proof of Lemma 24. With the aid of Equation (44), we obtain

$$D^{-\beta}[p_K(t,D)[D^\beta \tilde{w}(t)]] = D^{-\beta}[t \cdot D^2 + (c - bt) \cdot D - ab][D^\beta \tilde{w}(t)]$$

$$= [(t \cdot D - \beta)D + c \cdot D - b(t \cdot D - \beta) - ab]\tilde{w}(t) \tag{85}$$

$$= [t \cdot D^2 + (c - \beta - bt) \cdot D - (a - \beta)b]\tilde{w}(t) = p_{\tilde{K}}(t,D)\tilde{w}(t),$$

which gives Equation (73). When $\tilde{u}(t) = D^\beta \tilde{w}(t)$, Equation (64) shows that the lhs of Equation (85) is equal to $D^{-\beta}[\tilde{f}(t) + u_0(c-1)\delta(t)]$ and hence Equation (85) gives Equation (75). $\square$

Proof of Theorem 2. Theorem 1 states that when $w_g(t)$ is given by Equation (77), $\tilde{w}_g(t) := w_g(t)\tilde{H}(t)$ is the particular solution of Equation (75) for the term $f_\beta(t)\tilde{H}(t)$, and Lemma 24 states that $\tilde{u}_f(t) = D^\beta \tilde{w}_g(t)$ is the particular solution of Equation (64) for the term $f(t)\tilde{H}(t)$. Lemma 7 shows that $\tilde{u}_f(t) = u_f(t)\tilde{H}(t)$ if $u_f(t)$ is given by $u_f(t) = {}_0D_R^\beta w_g(t)$, since $\beta \notin \mathbb{Z}_{>-1}$. $\square$

5. Particular Solution of the Hypergeometric Differential Equation

Let

$$p_H(t,s) := t(1-t) \cdot s^2 + (c - (a+b+1)t) \cdot s - ab, \tag{86}$$

where $a \in \mathbb{C}$, $b \in \mathbb{C}$ and $c \in \mathbb{C}$ are constants. Then the hypergeometric differential equation with an inhomogeneous term $f(t)$ is given by

$$p_H(t, \frac{d}{dt})u(t) := t(1-t) \cdot \frac{d^2}{dt^2}u(t) + (c - (a+b+1)t) \cdot \frac{d}{dt}u(t) - ab \cdot u(t) = f(t), \quad t > 0. \tag{87}$$

If $f(t) = 0$ and $c \notin \mathbb{Z}$, the basic solutions of Equation (87) in [3,19] are given by

$$H_1(t) := {}_2F_1(a,b;c;t), \tag{88}$$

$$H_2(t) := t^{1-c} \cdot {}_2F_1(1+a-c, 1+b-c; 2-c; t), \tag{89}$$

where ${}_2F_1(a,b;c;z) = \sum_{k=0}^{\infty} \frac{(a)_k(b)_k}{k!(c)_k}z^k$ of $z \in \mathbb{C}$ is the hypergeometric series.

We now obtain a particular solution of this equation by the method stated in Section 3 and used in Section 4.1.

5.1. Green's Function for the Hypergeometric Differential Equation when Condition 2(i) Is Satisfied

The following two lemmas are proved in Section 5.4.

Lemma 28. *Let $u(t)$ be the solution of Equation (87), which is expressed by Equation (29). Then the differential equation satisfied by $\tilde{u}(t) = u(t)\tilde{H}(t)$ is given by*

$$p_H(t, D)\tilde{u}(t) = f(t)\tilde{H}(t) + u_0(c-1)\delta(t). \tag{90}$$

Lemma 29. *Let $H_1(t)$ and $H_2(t)$ be given by Equations (88) and (89), and $\psi_H(t, \tau)$ for fixed $\tau > 0$ be given by*

$$\psi_H(t, \tau) := c_1(\tau) \cdot H_1(t) + c_2(\tau) \cdot H_2(t), \tag{91}$$

where $c_1(\tau)$ and $c_2(\tau)$ are constants which are so chosen that $\psi_H(t, \tau) = 0$ when $t = \tau$. Then $G_H(t, 0)$ and $G_H(t, \tau)$ given by

$$G_H(t, 0) = \frac{1}{c-1} \cdot H_1(t)H(t), \quad G_H(t, \tau) = \frac{1}{\tau(1-\tau)\psi'_H(\tau, \tau)}\psi_H(t, \tau)H(t-\tau), \tag{92}$$

are the Green's functions, so that $\tilde{G}_H(t, 0) = G_H(t, 0)\tilde{H}(t)$ and $\tilde{G}_H(t, \tau) = G_H(t, \tau)\tilde{H}(t-\tau)$ satisfy

$$p_H(t, D)\tilde{G}_H(t, 0) = \delta(t), \quad p_H(t, D)\tilde{G}_H(t, \tau) = \delta(t-\tau). \tag{93}$$

Theorem 4. *Let $G_H(t, \tau)$ and $\psi_H(t, \tau)$ be those given in Lemma 29, Condition 2(i) be satisfied, and $u_f(t)$ be given by*

$$u_f(t) = \int_0^\infty G_H(t, \tau)f(\tau)d\tau = \int_0^t \psi_H(t, \tau)\frac{f(\tau)}{\tau(1-\tau)\psi'_H(\tau, \tau)}d\tau. \tag{94}$$

Then $u_f(t)$ and $u_f(t)\tilde{H}(t)$ are particular solutions of Equations (87) and (90) for the terms $f(t)$ and $f(t)\tilde{H}(t)$, respectively.

Proof. This is confirmed with the aid of Lemma 18. □

The condition $\psi_H(\tau, \tau) = 0$ requires that $c_1(\tau) \cdot H_1(\tau) = -c_2(\tau) \cdot H_2(\tau)$, and hence we may choose $\psi_H(t, \tau)$ as

$$\psi_H(t, \tau) = \begin{cases} H_1(t) - \frac{H_1(\tau)}{H_2(\tau)} \cdot H_2(t), & |H_1(\tau)| < |H_2(\tau)|, \\ \frac{H_2(\tau)}{H_1(\tau)} \cdot H_1(t) - H_2(t), & |H_1(\tau)| \geq |H_2(\tau)|. \end{cases} \tag{95}$$

5.2. Green's Function for the Hypergeometric Differential Equation in which Condition 2(ii) Is Satisfied

We give the solution of Equation (87) of which the inhomogeneous term $f(t)$ satisfies Condition 2(ii), so that $\tilde{f}(t) = D^\beta \tilde{f}_\beta(t)$, and $f_\beta(t)$ satisfies Condition 2(i).

The following lemma is proved in Section 5.4.

Lemma 30. *Let $\tilde{u}(t)$ be a solution of Equation (90), and $p_{\tilde{H}}(t, D)$ be related with $p_H(t, D)$ given by Equation (86), by*

$$p_{\tilde{H}}(t, D) = D^{-\beta}p_H(t, D)D^\beta. \tag{96}$$

Then

$$p_{\tilde{H}}(t, D) := t(1-t) \cdot D^2 + (c - \beta - (a+b-2\beta+1)t) \cdot D - (a-\beta)(b-\beta). \tag{97}$$

and $\tilde{w}(t) = D^{-\beta}\tilde{u}(t)$ satisfies

$$p_{\tilde{H}}(t,D)\tilde{w}(t) = f_\beta(t)\tilde{H}(t) + u_0(c-1)D^{-\beta}\delta(t). \tag{98}$$

Theorem 4 shows that the particular solution of Equation (98) for the term $f_\beta(t)\tilde{H}(t)$ is expressed by a particular soltion of

$$p_{\tilde{H}}(t,\frac{d}{dt})w(t) = f_\beta(t), \quad t > 0. \tag{99}$$

That solution is used in giving a solution of Equation (87) in Theorem 5 given below.

Remark 9. *We note that $p_{\tilde{H}}(t,D)$ given by Equation (97) is obtained from $p_H(t,D)$ given by Equation (86), by replacing a, b and c by $a-\beta$, $b-\beta$ and $c-\beta$, respectively, and hence the complementary solutions $H_{\beta,1}(t)$ and $H_{\beta,2}(t)$ and the Green's functions $G_{\tilde{H}}(t,0)$ and $G_{\tilde{H}}(t,\tau)$ of Equation (99) are obtained from those $H_1(t)$, $H_2(t)$, $G_H(t,0)$ and $G_H(t,\tau)$, respectively, of Equation (87), by the same replacement.*

Now in place of Theorem 4, we have the following theorem, whose proof is given in Section 5.4.

Theorem 5. *Let $G_{\tilde{H}}(t,\tau)$ and $\psi_{\tilde{H}}(t,\tau)$ be obtained from $G_H(t,\tau)$ and $\psi_H(t,\tau)$ by the replacement stated in Remark 9, Condition 2(ii) be satisfied, and $w_g(t)$ be given by*

$$w_g(t) = \int_0^t G_{\tilde{H}}(t,\tau)f_\beta(\tau)d\tau = \int_0^t \psi_{\tilde{H}}(t,\tau)\frac{f_\beta(\tau)}{\tau\psi'_{\tilde{H}}(\tau,\tau)}d\tau. \tag{100}$$

Then $u_f(t) := {}_0D_R^\beta w_g(t)$ and $u_f(t)\tilde{H}(t)$ are particular solutions of Equations (87) and (90) for the terms $f(t)$ and $f(t)\tilde{H}(t)$, respectively.

5.3. Particular Solution of the Hypergeometric Differential Equation in which Condition 2(iii) Is Satisfied

We give the solution of Equation (87) of which the inhomogeneous term $f(t)$ satisfies Condition 2(iii), so that $f(t) = g_\nu(t) = \frac{1}{\Gamma(\nu)}t^{\nu-1}$ and $\tilde{f}(t) = f(t)\tilde{H}(t) = \hat{f}(D)\delta(t) = \hat{g}_\nu(D)\delta(t) = D^{-\nu}\delta(t)$. Here we use $\beta \notin \mathbb{Z}_{>-1}$ in place of $-\nu$.

Lemma 31. *Let $\tilde{f}(t) = D^\beta\delta(t)$, $p_{\tilde{H}}(t,D)$ be given by Equation (97), and $\tilde{G}_{\tilde{H}}(t,0)$ satisfy*

$$p_{\tilde{H}}(t,D)\tilde{G}_{\tilde{H}}(t,0) = \delta(t). \tag{101}$$

Then the particular solution of Equation (90) for the term $f(t)\tilde{H}(t) = D^\beta\delta(t)$ is given by

$$\tilde{u}_f(t) = D^\beta\tilde{G}_{\tilde{H}}(t,0). \tag{102}$$

Proof. $\tilde{u}_f(t)$ satisfies $p_H(t,D)\tilde{u}_f(t) = D^\beta\delta(t)$ and hence $p_{\tilde{H}}(t,D)D^{-\beta}\tilde{u}_f(t) = \delta(t)$ by Equation (96). Comparing this with Equation (101), we see that Equation (102) is satisfied. □

By Remark 9, with the aid of $G_H(t,0)$ given by Equations (92) and (88), we have

Lemma 32. *As the solution of Equation (101), we obtain*

$$\tilde{G}_{\tilde{H}}(t,0) = G_{\tilde{H}}(t,0)\tilde{H}(t), \quad G_{\tilde{H}}(t,0) = \frac{1}{c-\beta-1}\cdot {}_2F_1(a-\beta,b-\beta;c-\beta;t). \tag{103}$$

Theorem 6. *Let Condition* 2*(iii) be satisfied, and* $G_{\tilde{H}}(t,0)$ *be given by Equation* (103). *Then the particular solution of Equation* (87) *is given by*

$$u_f(t) = {}_0D_R^\beta G_{\tilde{H}}(t,0) = \frac{t^{-\beta}}{(c-\beta-1)\Gamma(1-\beta)} \cdot {}_3F_2(1, a-\beta, b-\beta; 1-\beta, c-\beta; t), \tag{104}$$

where ${}_3F_2(a_1,a_2,a_3;c_1,c_2;z) = \sum_{k=0}^\infty \frac{(a_1)_k(a_2)_k(a_3)_k}{k!(c_1)_k(c_2)_k}z^k$.

Proof. $\tilde{u}_f(t)$ is given by Equations (102) and (103). Lemma 7 shows that $\tilde{u}_f(t) = u_f(t)\tilde{H}(t)$ if $u_f(t)$ is given by (104), since $\beta \notin \mathbb{Z}_{>-1}$. □

5.4. Proofs of Lemmas 28–30 *and Theorem* 5

Proof of Lemma 28. Equation (90) is a special one of (48), where p_n, n, $a_2(t)$, $a_1(t)$ and $a_0(t)$ stand for p_H, 2, $t(1-t)$, $c-(a+b+1)t$ and $-ab$, respectively. Hence by using Equation (37), we see that the second term on the rhs of Equation (48) becomes

$$\begin{aligned}\sum_{l=1}^2 a_l(t)\langle D^l\hat{u}(D)\rangle_0\delta(t) &= t(1-t)(u_0D+u_1)\delta(t) + (c-(a+b+1)t)u_0\delta(t) \\ &= u_0(D[t(1-t)\delta(t)] - (1-2t)\delta(t)) + cu_0\delta(t) = u_0(c-1)\delta(t).\end{aligned} \tag{105}$$

□

We prepare the following lemma.

Lemma 33. *Let* $\tau \geq 0$, $p_H(t,\frac{d}{dt})\psi(t)\cdot H(t-\tau) \in \mathcal{L}_{loc}^1(\mathbb{R})$ *and* $\tilde{\psi}_\tau(t) = \psi(t)\tilde{H}(t-\tau)$. *Then*

$$p_H(t,D)\tilde{\psi}_\tau(t) = p_H(t,\frac{d}{dt})\psi(t)\cdot\tilde{H}(t-\tau) + \sum_{l=1}^2 a_l(t)\langle D^l\tilde{\psi}_\tau(D)\rangle_0\delta(t-\tau), \tag{106}$$

where $a_2(t) = t(1-t)$, $a_1(t) = c-(a+b+1)t$, $a_0(t) = -ab$, *and*

$$\begin{aligned}\sum_{l=1}^2 a_l(t)\langle D^l\tilde{\psi}_\tau(D)\rangle_0\delta(t-\tau) &= \tau(1-\tau)\psi'(\tau)\delta(t-\tau) + \psi(\tau)(c-1-(a+b-1)\tau)\delta(t-\tau) \\ &\quad + \psi(\tau)\tau(1-\tau)D\delta(t-\tau).\end{aligned} \tag{107}$$

Proof. Equation (106) is a special one of Equation (18) in Corollary 3. The lhs of Equation (107) is evaluated with the aid of Equation (16) as follows:

$$\begin{aligned}\sum_{l=1}^2 a_l(t)\langle D^l\tilde{\psi}_\tau(D)\rangle_0\delta(t-\tau) &= t(1-t)[\psi'(\tau)+\psi(\tau)D]\delta(t-\tau) + (c-(a+b+1)t)\psi(\tau)\delta(t-\tau) \\ &= [t(1-t)\psi'(\tau)+\psi(\tau)(c-(a+b+1)t)]\delta(t-\tau) + \psi(\tau)t(1-t)D\delta(t-\tau) \\ &= \tau(1-\tau)\psi'(\tau)\delta(t-\tau) + \psi(\tau)(c-1-(a+b-1)\tau)\delta(t-\tau) + \psi(\tau)\tau(1-\tau)D\delta(t-\tau),\end{aligned}$$

where Equation (19) with $h(t)$ and $\tilde{u}(t)$ replaced by $t(1-t)$ and $\delta(t-\tau)$, respectively, is used in the last term of the third member. □

Proof of Lemma 29. If we put $\psi(t) = \frac{1}{c-1}\cdot H_1(t)$ and $\tau = 0$ in Equation (106), then we have $\tilde{\psi}_\tau(t) = \tilde{G}(t,0)$ on the lhs, and the rhs is $\psi(0)(c-1)\delta(t) = \delta(t)$, since $H_1(0) = 1$. If we put $\psi(t) = \frac{1}{\tau(1-\tau)\psi_H'(\tau,\tau)}\psi_H(t,\tau)$ and $\psi(\tau) = 0$ in Equation (106), then we have $\tilde{\psi}_\tau(t) = \tilde{G}(t,0)$ on the lhs, and the rhs is $\tau(1-\tau)\psi'(\tau)\delta(t-\tau) = \delta(t-\tau)$. □

Proof of Lemma 30. With the aid of Equations (44) and (45), we obtain

$$
\begin{aligned}
D^{-\beta}[p_H(t,D)[D^{\beta}\tilde{w}(t)]] &= D^{-\beta}[t(1-t)\cdot D^2 + (c-(a+b+1)t)\cdot D - ab][D^{\beta}\tilde{w}(t)] \\
&= [(t\cdot D-\beta)D - (t^2\cdot D^2 - 2\beta tD + \beta(\beta+1)) + c\cdot D - (a+b+1)(t\cdot D-\beta) - ab]\tilde{w}(t) \\
&= [t(1-t)\cdot D^2 + (c-\beta-(a+b-2\beta+1)t)\cdot D - (a-\beta)(b-\beta)]\tilde{w}(t) = p_{\tilde{H}}(t,D)\tilde{w}(t),
\end{aligned}
\tag{108}
$$

which gives Equation (96). When $\tilde{u}(t) = D^{\beta}\tilde{w}(t)$, Equation (90) shows that the lhs of (108) is equal to $D^{-\beta}[\tilde{f}(t) + u_0(c-1)\delta(t)]$ and hence Equation (108) gives Equation (98). □

Proof of Theorem 5. Theorem 4 states that when $w_g(t)$ is given by Equation (100), $\tilde{w}_g(t) := w_g(t)\tilde{H}(t)$ is the particular solution of Equation (98) for the term $f_\beta(t)\tilde{H}(t)$, and Lemma 30 states that $\tilde{u}_f(t) = D^{\beta}\tilde{w}_g(t)$ is the particular solution of Equation (90) for the term $f(t)\tilde{H}(t)$. Lemma 7 shows that $\tilde{u}_f(t) = u_f(t)\tilde{H}(t)$ if $u_f(t)$ is given by $u_f(t) = {}_0D_R^{\beta}w_g(t)$, since $\beta \notin \mathbb{Z}_{>-1}$. □

6. Solution of Inhomogeneous Differential Equations with Constant Coefficients

In Section 3 we discuss the solution of an inhomogeneous differential equation with constant coefficients, which takes the form of Equation (56), in terms of the Green's function and distribution theory. In this and next sections, we discuss it in terms of the Green's function and the Laplace transform.

Lemma 34. *Let $f(t)$ have the AC-Laplace transform $\hat{f}(s) = \mathcal{L}_H[f(t)]$. Then the solution $u(t)$ of Equation (56) has the AC-Laplace transform $\hat{u}(s) = \mathcal{L}_H[u(t)]$, which satisfies*

$$
p_n(s)\hat{u}(s) := \sum_{l=0}^{n} a_l s^l \hat{u}(s) = \hat{f}(s) + \sum_{l=1}^{n} a_l \langle s^l \hat{u}(s) \rangle_0,
\tag{109}
$$

where $\langle s^l \hat{u}(s) \rangle_0$ are given by Equations (36) and (37) with D replaced by s.

Proof. This is confirmed with the aid of Lemma 12. □

We introduce the Green's function $G(t)$ so that its Laplace transform $\hat{G}(s)$ satisfies

$$
p_n(s)\hat{G}(s) = 1,
\tag{110}
$$

and hence $\hat{G}(s) = \frac{1}{p_n(s)}$. Multiplying this to Equation (109), we obtain

$$
\hat{u}(s) = \hat{G}(s)\hat{f}(s) + \hat{G}(s)\sum_{l=1}^{n} a_l \langle s^l \hat{u}(s) \rangle_0.
\tag{111}
$$

Comparing Equations (110) and (109), we see that the differential equation for the Green's function $G(t)$, whose Laplace transform $\hat{G}(s)$ satisfies Equation (110), is Equation (60), and the initial values of $G(t)$ and its derivatives satisfy $\langle p(s)\hat{G}(s) \rangle_0 = 1$, and hence are given by Equation (61). Thus we confirm Lemma 20.

By the inverse Laplace transform of Equation (111), we obtain Lemma 21.

6.1. Solution of an Inhomogeneous Differential Equation of the First Order

We consider an inhomogeneous differential equation of the first order:

$$
p_1\left(\frac{d}{dt}\right)u(t) := \frac{d}{dt}u(t) + cu(t) = f(t), \quad t > 0,
\tag{112}
$$

where $c \in \mathbb{C}$ is a constant.

By Lemma 34, we obtain the following equation for $\hat{u}(s) = \mathcal{L}_H[u(t)]$:

$$p_1(s)\hat{u}(s) = (s+c)\hat{u}(s) = u_0 + \hat{f}(s). \tag{113}$$

Following Section 6, we introduce the Green's function $G(t)$, so that its Laplace transform $\hat{G}(s)$, which satisfies Equation (110), is given by $p_1(s)\hat{G}(s) = (s+c) \cdot \hat{G}(s) = 1$, and hence we have

$$\hat{G}(s) = \frac{1}{p_1(s)} = \frac{1}{s+c}. \tag{114}$$

By using this equation in Equation (113) and putting $\hat{u}_f(s) = \hat{G}(s)\hat{f}(s)$, we obtain

$$\hat{u}(s) = u_0\hat{G}(s) + \hat{u}_f(s), \quad \hat{u}_f(s) = \hat{G}(s)\hat{f}(s). \tag{115}$$

By the inverse Laplace transform of Equation (114), we obtain

$$G(t) = e^{-ct}H(t). \tag{116}$$

Theorem 7. *Let Condition 2(i) or 2(ii) be satisfied. Then the solution of Equation (112) is given by*

$$u(t) = u_0G(t) + u_f(t), \quad t > 0, \tag{117}$$

where

$$u_f(t) = \int_0^t G(t-\tau)f(\tau)d\tau = e^{-ct}\int_0^t e^{c\tau}f(\tau)d\tau, \quad t > 0, \tag{118}$$

or

$$u_f(t) = {}_0D_R^\beta\left[\int_0^t G(t-\tau)f_\beta(\tau)d\tau\right] = {}_0D_R^\beta\left[e^{-ct}\int_0^t e^{c\tau}f_\beta(\tau)d\tau\right], \quad t > 0, \tag{119}$$

according as Condition 2(i) or 2(ii) is satisfied.

Proof. By the AC-Laplace transform of Equation (117), we obtain Equation (115). When Condition 2(i) is satisfied, the Laplace transform of Equation (118) is $\hat{u}_f(s) = \hat{G}(s)\hat{f}(s)$. When Condition 2(ii) is satisfied, we confirm that the AC-Laplace transform of Equation (119) is $\hat{u}_f(s) = s^\beta\hat{G}(s)\hat{f}_\beta(s)$, with the aid of Equations (28) and (27). $\square$

Theorem 8. *Let Condition 2(iii) be satisfied, so that $\hat{f}(s) = s^\beta$ for $\beta \in \mathbb{C}\backslash\mathbb{Z}_{>-1}$. Then the solution of Equation (112) is given by Equation (117) with*

$$u_f(t) = t^{-\beta}\sum_{k=0}^\infty \frac{(1)_k(-ct)^k}{k!\Gamma(k+1-\beta)} = \frac{t^{-\beta}}{\Gamma(1-\beta)} \cdot {}_1F_1(1;1-\beta;-ct). \tag{120}$$

Proof. By using Equation (114) in Equation (115), we have

$$\hat{u}_f(s) = \frac{s^\beta}{s+c} = \sum_{k=0}^\infty (-c)^k s^{-k-1+\beta}. \tag{121}$$

By the inverse Laplace transform of this equation, we obtain Equation (120), where we use $\Gamma(k+1-\beta) = \Gamma(1-\beta)(1-\beta)_k$. $\square$

6.2. Solution of an Inhomogeneous Differential Equation of the Second Order

We consider an inhomogeneous differential equation of the second order:

$$p_2(\frac{d}{dt})u(t) := \frac{d^2}{dt^2}u(t) + b\frac{d}{dt}u(t) + cu(t) = f(t), \quad t > 0, \tag{122}$$

where $b \in \mathbb{C}$ and $c \in \mathbb{C}$ are constants.

By Lemma 34, we obtain the following equation for $\hat{u}(s) = \mathcal{L}_H[u(t)]$:

$$p_2(s)\hat{u}(s) := (s^2 + bs + c)\hat{u}(s) = (u_0 s + u_1) + bu_0 + \hat{f}(s). \tag{123}$$

Following Section 6, we introduce the Green's function $G(t)$, so that its Laplace transform $\hat{G}(s)$, which satisfies Equation (110), is given by $p_2(s)\hat{G}(s) = (s^2 + bs + c) \cdot \hat{G}(s) = 1$, and hence we have

$$\hat{G}(s) = \frac{1}{p_2(s)} = \frac{1}{s^2 + bs + c}. \tag{124}$$

By using Equation (124) in Equation (123), we obtain

$$\hat{u}(s) = u_0 s\hat{G}(s) + (u_1 + bu_0)\hat{G}(s) + \hat{u}_f(s), \tag{125}$$

where

$$\hat{u}_f(s) = \hat{G}(s)\hat{f}(s). \tag{126}$$

Equation:

$$p_2(s) := s^2 + bs + c = 0, \tag{127}$$

has one or two roots according as $c = \frac{b^2}{4}$ or not. If $c \neq \frac{b^2}{4}$, then Equation (127) has two different roots μ_1 and μ_2, so that $b = -(\mu_1 + \mu_2)$ and $c = \mu_1\mu_2$, and Equation (124) gives

$$\hat{G}(s) = \frac{1}{(s - \mu_1)(s - \mu_2)} = \frac{1}{\mu_1 - \mu_2}(\frac{1}{s - \mu_1} - \frac{1}{s - \mu_2}). \tag{128}$$

By the inverse Laplace transform, we then obtain

$$G(t) = \frac{1}{\mu_1 - \mu_2}(e^{\mu_1 t} - e^{\mu_2 t})H(t). \tag{129}$$

If $c = \frac{b^2}{4}$, then Equation (127) has only one root μ_1, so that $b = -2\mu_1$ and $c = \mu_1^2$, and in place of Equation (128), we have

$$\hat{G}(s) = \frac{1}{(s - \mu_1)^2}. \tag{130}$$

By the inverse Laplace transform, we then obtain

$$G(t) = te^{\mu_1 t}H(t). \tag{131}$$

Theorem 9. *Let Condition 2(i) or 2(ii) be satisfied. Then the solution of Equation (122) is given by*

$$u(t) = u_0 \frac{d}{dt}G(t) + (u_1 + bu_0)G(t) + u_f(t), \quad t > 0, \tag{132}$$

where

$$u_f(t) = \int_0^t G(t-\tau)f(\tau)d\tau, \quad t > 0,$$ (133)

or

$$u_f(t) = {}_0D_R^\beta \left[\int_0^t G(t-\tau)f_\beta(\tau)d\tau \right], \quad t > 0,$$ (134)

according as Condition 2*(i) or* 2*(ii) is satisfied. If* $c \neq \frac{b^2}{4}$, *then* $b = -(\mu_1 + \mu_2)$, $c = \mu_1\mu_2$, *and* $G(t)$ *given by Equation* (129) *is used in Equation* (132). *If* $c = \frac{b^2}{4}$, *then* $b = -2\mu_1$, $c = \mu_1^2$, *and* $G(t)$ *given by Equation* (131) *is used there.*

Proof. By the AC-Laplace transform of Equation (132), we obtain Equation (125). When Condition 2(ii) is satisfied, we have $\hat{f}(s) = s^\beta \hat{f}_\beta(s)$. By using Equations (28) and (27), we confirm that the AC-Laplace transform of $u_f(t)$ given by Equation (134) is $s^\beta \hat{G}(s)\hat{f}_\beta(s)$. $\square$

Theorem 10. *Let Condition* 2*(iii) be satisfied, be satisfied, so that* $\hat{f}(s) = s^\beta$ *for* $\beta \in \mathbb{C}\backslash\mathbb{Z}_{>-1}$. *Then the solution of Equation* (122) *is given by Equation* (132) *with* $u_f(t)$ *given as follows. If* $c \neq \frac{b^2}{4}$, *then* $b = -(\mu_1 + \mu_2)$, $c = \mu_1\mu_2$, *and*

$$u_f(t) = \frac{1}{\mu_1 - \mu_2}\frac{t^{-\beta}}{\Gamma(1-\beta)}[{}_1F_1(1; 1-\beta; \mu_1 t) - {}_1F_1(1; 1-\beta; \mu_2 t)].$$ (135)

If $c = \frac{b^2}{4}$, *then* $b = -2\mu_1$, $c = \mu_1^2$, *and*

$$u_f(t) = \frac{t^{1-\beta}}{\Gamma(1-\beta)} \cdot {}_1F_1(2; 1-\beta; \mu_1 t).$$ (136)

Proof. When $c \neq \frac{b^2}{4}$, by using Equation (128) in Equation (126), we have

$$\hat{u}_f(s) = \frac{1}{\mu_1 - \mu_2}\left(\frac{s^\beta}{s - \mu_1} - \frac{s^\beta}{s - \mu_2}\right).$$ (137)

In the proof of Theorem 8, we obtain Equation (120), by the inverse Laplace transform of $\hat{u}_f(s)$ given by Equation (121). By the corresponding inverse Laplace transform of Equation (137), we obtain Equation (135). When $c = \frac{b^2}{4}$, by using Equation (130), we have

$$\hat{u}_f(s) = -s^\beta \frac{d}{ds}\frac{1}{s - \mu_1} = -s^\beta \frac{d}{ds}\sum_{k=0}^\infty \mu_1^k s^{-k-1} = s^\beta \sum_{k=0}^\infty \mu_1^k(k+1)s^{-k-2}.$$ (138)

By the inverse Laplace transform, we have

$$u_f(t) = t^{1-\beta}\sum_{k=0}^\infty \frac{(k+1)(\mu_1 t)^k}{\Gamma(k+1-\beta)} = \frac{t^{1-\beta}}{\Gamma(1-\beta)}\sum_{k=0}^\infty \frac{(2)_k(\mu_1 t)^k}{k!(1-\beta)_k},$$ (139)

which gives Equation (136). $\square$

6.3. Application of the Theorems in Section 6.1

We consider an inhomogeneous differential equation with polynomial coefficients of the first order:

$$\frac{d}{dx}w(x) - 2xw(x) = 2xg(x) = 1, \quad x \in \mathbb{R}.$$ (140)

We put $t = x^2$, $u(t) = u(x^2) = w(x)$ and $f(t) = f(x^2) = g(x)$. Then $u(t)$ satisfies

$$\frac{d}{dt}u(t) - u(t) = f(t), \quad t \in \mathbb{R}_{>0}, \tag{141}$$

where

$$f(t) = \frac{1}{2x} = \frac{t^{-1/2}}{2} = \frac{\Gamma(1/2)}{2}g_{1/2}(t) = \frac{\sqrt{\pi}}{2}g_{1/2}(t), \tag{142}$$

where $g_{1/2}(t)$ is defined by Equation (26).

Lemma 35. *Let $u(t)$ be a solution of Equation (141). Then $w(x) = u(x^2)$ gives a solution of Equation (140).*

Lemma 36. *The solution of Equation (141) is given by Equation (117), where $u_0 = u(0)$ and*

$$G(t) = e^t, \quad u_f(t) = \frac{\sqrt{\pi}}{2}\sum_{k=0}^{\infty}\frac{t^{k+1/2}}{\Gamma(k+\frac{3}{2})}, \quad t \in \mathbb{R}_{>0}. \tag{143}$$

The solution of Equation (140) is given by

$$w(x) = w_0 G_w(x) + w_g(x), \quad x \in \mathbb{R}, \tag{144}$$

where $w_0 = w(0)$ and

$$G_w(x) = e^{x^2}, \quad w_g(x) = \frac{\sqrt{\pi}}{2}\sum_{k=0}^{\infty}\frac{x^{2k+1}}{\Gamma(k+\frac{3}{2})}, \quad x \in \mathbb{R}. \tag{145}$$

Proof. By using $c = -1$, $\beta = \frac{1}{2}$, Equation (142) in Equations (116) and (120), we see that the solution of Equation (141) is given by Equation (117) with Equation (143). Now the solution of Equation (140) is obtained from it with the aid of Lemma 35. □

Lemma 37. *The asymptotic behavior of $u_f(t)$ and $w_g(x)$ are given by*

$$u_f(t) = \frac{\sqrt{\pi}}{2}e^t - \sum_{k=0}^{\infty}\frac{(-1)^k(-\frac{1}{2})_k}{2t^{k+1/2}}, \quad t \to \infty, \tag{146}$$

$$w_g(x) = \begin{cases} \frac{\sqrt{\pi}}{2}e^{x^2} - \sum_{k=0}^{\infty}\frac{(-1)^k(-\frac{1}{2})_k}{2x^{2k+1}}, & x \to \infty, \\ -\frac{\sqrt{\pi}}{2}e^{x^2} - \sum_{k=0}^{\infty}\frac{(-1)^k(-\frac{1}{2})_k}{2x^{2k+1}}, & x \to -\infty. \end{cases} \tag{147}$$

Proof. By using Equations (142) and (143) in Equation (118), we have

$$u_f(t) = \int_0^t e^{t-\tau}\frac{\tau^{-1/2}}{2}d\tau = e^t\frac{\Gamma(\frac{1}{2})}{2} - \frac{1}{2}\int_0^{\infty}e^{-\tau}(t+\tau)^{1/2}d\tau = e^t\frac{\sqrt{\pi}}{2} - \frac{1}{2}\sum_{k=0}^{\infty}\frac{(-1)^k(-\frac{1}{2})_k}{t^{k+1/2}}, \tag{148}$$

which gives Equation (146). Equation (147) is obtained from it with the aid of Lemma 35. □

Equation (145) shows that the particular solution $w_g(x)$ of Equation (140) is an odd function of x. As a consequence of this fact, the asymptotic behavior of $w_g(x)$ is given by Equation (147).

Acknowledgments: The authors are grateful to the reviewers of this paper. Following their suggestions, the last paragraph of Introduction and proofs of some theorems are included, and proofs of some lemmas are rewritten in details.

Author Contributions: In the last paragraph of Introduction, recent study on the solution of differential equations is reviewed. After the paper [2] was published, Tohru Morita wrote a manuscript giving particular solutions of

Kummer's and the hypergeometric differentail equations in terms of distribution theory and the Green's function. Tohru Morita and Ken-ichi Sato worked together to improve the manuscript to the present paper, where the solution of differential equations with constant coefficients is written in the form where the role of the Green's function is seen in the framework of the AC-Laplace transform.

Conflicts of Interest: The authors declare no conflict of interest.

References

1. Morita, T.; Sato, K. Solution of Differential Equations with the Aid of Analytic Continuation of Laplace Transform. *Appl. Math.* **2014**, *5*, 1309–1319.
2. Morita, T.; Sato, K. Solution of Differential Equations of Polynomial Coefficients with the Aid of Analytic Continuation of Laplace Transform. *Mathematics* **2016**, *4*, 19.
3. Abramowitz, M.; Stegun, I.A. *Handbook of Mathematical Functions with Formulas, Graphs and Mathematical Tables*; Dover Publ., Inc.: New York, NY, USA, 1972; Chapter 13.
4. Morita, T.; Sato, K. Asymptotic Expansions of Fractional Derivatives and Their Applications. *Mathematics* **2015**, *3*, 171–189.
5. Zemanian, A.H. *Distribution Theory and Transform Analysis*; Dover Publ., Inc.: New York, NY, USA, 1965.
6. Godoy, E.; Ronveaux, A.; Zarzo, A.; Area, I. Connection Problems for Polynomial Solutions of Nonhomogeneous Differential and Difference Equations. *J. Comput. Appl. Math.* **1998**, *99*, 177–187.
7. Morita, T.; Sato, K. Remarks on the Solution of Laplace's Differential Equation and Fractional Differential Equation of That Type. *Appl. Math.* **2013**, *4*, 13–21.
8. Morita, T.; Sato, K. Solution of Laplace's Differential Equation and Fractional Differential Equation of That Type. *Appl. Math.* **2013**, *4*, 26–36.
9. Yosida, K. *Operational Calculus*; Springer: New York, NY, USA, 1982; Chapter VII.
10. Yosida, K. The Algebraic Derivative and Laplace's Differential Equation. *Proc. Jpn. Acad. Ser. A* **1983**, *59*, 1–4.
11. Morita, T.; Sato, K. Solution of Fractional Differential Equation in Terms of Distribution Theory. *Interdiscip. Inf. Sci.* **2006**, *12*, 71–83.
12. Morita, T.; Sato, K. Neumann-Series Solution of Fractional Differential Equation. *Interdiscip. Inf. Sci.* **2010**, *16*, 127–137.
13. Gelfand, I.M.; Shilov, G.E. *Generalized Functions*; Academic Press Inc.: New York, NY, USA, 1964; Volume 1.
14. Vladimirov, V.S. *Methods of the Theory of Generalized Functions*; Taylor & Francis Inc.: New York, NY, USA, 2002.
15. Schwartz, L. *Théorie des Distributions*; Hermann: Paris, France, 1966.
16. Morita, T.; Sato, K. Liouville and Riemann-Liouville Fractional Derivatives via Contour Integrals. *Fract. Calc. Appl. Anal.* **2013**, *16*, 630–653.
17. Podlubny, I. *Fractional Differential Equations*; Academic Press: San Diego, CA, USA, 1999.
18. Miller, K.S.; Ross, B. *An Introduction to the Fractional Calculus and Fractional Differential Equations*; John Wiley: New York, NY, USA, 1993.
19. Magnus, M.; Oberhettinger, F. *Formulas and Theorems for the Functions of Mathematical Physics*; Chelsea Publ. Co.: New York, NY, USA, 1949; Chapter VI.

![Σ mathematics logo] *mathematics*

MDPI

Article

Babenko's Approach to Abel's Integral Equations

Chenkuan Li * and Kyle Clarkson

Department of Mathematics and Computer Science, Brandon University, Brandon, MB R7A 6A9, Canada;
kyleclarkson17@hotmail.com
* Correspondence: lic@brandonu.ca

Received: 25 January 2018; Accepted: 18 February 2018; Published: 1 March 2018

Abstract: The goal of this paper is to investigate the following Abel's integral equation of the second kind: $y(t) + \frac{\lambda}{\Gamma(\alpha)} \int_0^t (t-\tau)^{\alpha-1} y(\tau) d\tau = f(t)$, $(t > 0)$ and its variants by fractional calculus. Applying Babenko's approach and fractional integrals, we provide a general method for solving Abel's integral equation and others with a demonstration of different types of examples by showing convergence of series. In particular, we extend this equation to a distributional space for any arbitrary $\alpha \in R$ by fractional operations of generalized functions for the first time and obtain several new and interesting results that cannot be realized in the classical sense or by the Laplace transform.

Keywords: distribution; fractional calculus; convolution; series convergence; Laplace transform; Gamma function; Mittag–Leffler function

JEL Classification: 46F10; 26A33; 45E10; 45G05

1. Introduction

Abel's equations are related to a wide range of physical problems, such as heat transfer [1], nonlinear diffusion [2], the propagation of nonlinear waves [3], and applications in the theory of neutron transport and traffic theory. There have been many approaches including numerical analysis thus far to studying Abel's integral equations as well as their variants with many applications [4–13]. In 1930, Tamarkin [14] discussed integrable solutions of Abel's integral equation under certain conditions by several integral operators. Sumner [15] studied Abel's integral equation from the point of view of the convolutional transform. Minerbo and Levy [16] investigated a numerical solution of Abel's integral equation using orthogonal polynomials. In 1985, Hatcher [17] worked on a nonlinear Hilbert problem of power type, solved in closed form by representing a sectionally holomorphic function by means of an integral with power kernel, and transformed the problem to one of solving a generalized Abel's integral equation. Singh et al. [18] obtained a stable numerical solution of Abel's integral equation using an almost Bernstein operational matrix. Recently, Li and Zhao [19] used the inverse of Mikusinski's operator of fractional order, based on Mikusinski's convolution, to construct the solution of the integral equation of Abel's type. Jahanshahi et al. [20] solved Abel's integral equation numerically on the basis of approximations of fractional integrals and Caputo derivatives. Saleh et al. [21] investigated the numerical solution of Abel's integral equation by Chebyshev polynomials. Kumar et al. [22] proposed a new and simple algorithm for Abel's integral equation, namely, the homotopy perturbation transform method (HPTM), based on the Laplace transform algorithm, and made the calculation for approximate solutions much easier. The main advance of this proposal is its capability of obtaining rapid convergent series for singular integral equations of Abel type. Recently, Li et al. [23] studied the following Abel's integral equation:

$$g(x) = \frac{1}{\Gamma(1-\alpha)} \int_0^x \frac{f(\zeta)}{(x-\zeta)^\alpha} d\zeta, \quad \text{where } \alpha \in R,$$

and its variants in the distributional (Schwartz) sense based on fractional calculus of distributions and derived new results that are not achievable in the classical sense.

The current work is grouped as follows. In Section 2, we briefly introduce the necessary concepts and definitions of fractional calculus of distributions in $\mathcal{D}'(R^+)$, which is described in Section 4. In Section 3, we solve Abel's integral equation of the second kind for $\alpha > 0$ and its variants by Babenko's approach, as well as fractional integrals with different types of illustrative examples. Often we obtain an infinite series as the solution of Abel's integral equation and then show its convergence. In Section 4, we extend Abel's integral equation into the distributional space $\mathcal{D}'(R^+)$ for all $\alpha \in R$ by a new technique of computing fractional operations of distributions. We produce some novel results that cannot be derived in the ordinary sense.

2. Fractional Calculus of Distributions in $\mathcal{D}'(R^+)$

In order to investigate Abel's integral equation in the generalized sense, we introduce the following basic concepts in detail. We let $\mathcal{D}(R)$ be the Schwartz space [24] of infinitely differentiable functions with compact support in R and $\mathcal{D}'(R)$ be the space of distributions defined on $\mathcal{D}(R)$. Further, we define a sequence $\phi_1(x)$, $\phi_2(x)$, $\cdots$, $\phi_n(x)$, $\cdots$, which converges to zero in $\mathcal{D}(R)$ if all these functions vanish outside a certain fixed bounded interval and converges uniformly to zero (in the usual sense) together with their derivatives of any order. The functional $\delta(x)$ is defined as

$$(\delta, \phi) = \phi(0),$$

where $\phi \in \mathcal{D}(R)$. Clearly, δ is a linear and continuous functional on $\mathcal{D}(R)$, and hence $\delta \in \mathcal{D}'(R)$. We let $\mathcal{D}'(R^+)$ be the subspace of $\mathcal{D}'(R)$ with support contained in R^+.

We define

$$\theta(x) = \begin{cases} 1 & \text{if } x > 0, \\ 0 & \text{if } x < 0. \end{cases}$$

Then

$$(\theta(x), \phi(x)) = \int_0^\infty \phi(x)dx \quad \text{for} \quad \phi \in \mathcal{D}(R),$$

which implies $\theta(x) \in \mathcal{D}'(R)$.

We let $f \in \mathcal{D}'(R)$. The distributional derivative of f, denoted by f' or df/dx, is defined as

$$(f', \phi) = -(f, \phi')$$

for $\phi \in \mathcal{D}(R)$.

Clearly, $f' \in \mathcal{D}'(R)$ and every distribution has a derivative. As an example, we show that $\theta'(x) = \delta(x)$, although $\theta(x)$ is not defined at $x = 0$. Indeed,

$$(\theta'(x), \phi(x)) = -(\theta(x), \phi'(x)) = -\int_0^\infty \phi'(x)dx = \phi(0) = (\delta(x), \phi(x)),$$

which claims

$$\theta'(x) = \delta(x).$$

It can be shown that the ordinary rules of differentiation apply also to distributions. For instance, the derivative of a sum is the sum of the derivatives, and a constant can be commuted with the derivative operator.

It follows from [24–26] that $\Phi_\lambda = \dfrac{x_+^{\lambda-1}}{\Gamma(\lambda)} \in \mathcal{D}'(R^+)$ is an entire function of λ on the complex plane, and

$$\left. \frac{x_+^{\lambda-1}}{\Gamma(\lambda)} \right|_{\lambda=-n} = \delta^{(n)}(x) \quad \text{for} \quad n = 0, 1, 2, \cdots \tag{1}$$

Clearly, the Laplace transform of Φ_λ is given by

$$\mathcal{L}\{\Phi_\lambda(x)\} = \int_0^\infty e^{-sx}\Phi_\lambda(x)dx = \frac{1}{s^\lambda}, \quad \text{Re}\lambda > 0, \quad \text{Re}s > 0,$$

which plays an important role in solving integral equations [27,28].

For the functional $\Phi_\lambda = \frac{x_+^{\lambda-1}}{\Gamma(\lambda)}$, the derivative formula is simpler than that for x_+^λ. In fact,

$$\frac{d}{dx}\Phi_\lambda = \frac{d}{dx}\frac{x_+^{\lambda-1}}{\Gamma(\lambda)} = \frac{(\lambda-1)x_+^{\lambda-2}}{\Gamma(\lambda)} = \frac{x_+^{\lambda-2}}{\Gamma(\lambda-1)} = \Phi_{\lambda-1}. \tag{2}$$

The convolution of certain pairs of distributions is usually defined as follows (see Gel'fand and Shilov [24], for example):

Definition 1. *Let f and g be distributions in $\mathcal{D}'(R)$ satisfying either of the following conditions:*

(a)   *Either f or g has bounded support (set of all essential points), or*
(b)   *the supports of f and g are bounded on the same side.*

Then the convolution $f * g$ is defined by the equation

$$((f * g)(x), \phi(x)) = (g(x), (f(y), \phi(x+y)))$$

for $\phi \in \mathcal{D}(R)$.

The classical definition of the convolution is as follows:

Definition 2. *If f and g are locally integrable functions, then the convolution $f * g$ is defined by*

$$(f * g)(x) = \int_{-\infty}^\infty f(t)g(x-t)dt = \int_{-\infty}^\infty f(x-t)g(t)dt$$

for all x for which the integrals exist.

We note that if f and g are locally integrable functions satisfying either of the conditions (a) or (b) in Definition 1, then Definition 1 is in agreement with Definition 2. It also follows that if the convolution $f * g$ exists by Definition 1 or 2, then the following equations hold:

$$f * g = g * f \tag{3}$$
$$(f * g)' = f * g' = f' * g \tag{4}$$

where all the derivatives above are in the distributional sense.

We let λ and μ be arbitrary complex numbers. Then it is easy to show that

$$\Phi_\lambda * \Phi_\mu = \Phi_{\lambda+\mu} \tag{5}$$

by Equation (2), without any help of analytic continuation mentioned in all current books.

We let λ be an arbitrary complex number and $g(x)$ be the distribution concentrated on $x \geq 0$. We define the primitive of order λ of g as a convolution in the distributional sense:

$$g_\lambda(x) = g(x) * \frac{x_+^{\lambda-1}}{\Gamma(\lambda)} = g(x) * \Phi_\lambda. \tag{6}$$

We note that the convolution on the right-hand side is well defined, as supports of g and Φ_λ are bounded on the same side.

Thus Equation (6) with various λ will not only give the fractional derivatives but also the fractional integrals of $g(x) \in \mathcal{D}'(R^+)$ when $\lambda \notin Z$, and it reduces to integer-order derivatives or integrals when $\lambda \in Z$. We define the convolution

$$g_{-\lambda} = g(x) * \Phi_{-\lambda}$$

as the fractional derivative of the distribution $g(x)$ with order λ, writing it as

$$g_{-\lambda} = \frac{d^\lambda}{dx^\lambda} g$$

for $\mathrm{Re}\lambda \geq 0$. Similarly, $\dfrac{d^\lambda}{dx^\lambda} g$ is interpreted as the fractional integral if $\mathrm{Re}\lambda < 0$.

In 1996, Matignon [26] studied fractional derivatives in the sense of distributions for fractional differential equations with applications to control processing and defined the fractional derivative of order λ of a continuous causal (zero for $t < 0$) function g as

$$\frac{d^\lambda}{dx^\lambda} g = g(x) * \Phi_{-\lambda},$$

which is a special case of Equation (6), as g belongs to $\mathcal{D}'(R^+)$. A very similar definition for the fractional derivatives of causal functions (not necessarily continuous) is given by Mainardi in [27].

As an example of finding a fractional derivative of a distribution, we let $g(x) \in \mathcal{D}'(R^+)$ be given by

$$g(x) = \begin{cases} 1 & \text{if } x \text{ is irrational and positive,} \\ 0 & \text{otherwise.} \end{cases}$$

Then the ordinary derivative of $g(x)$ does not exist. However, the distributional derivative of $g(x)$ does exist, and $g'(x) = \delta(x)$ on the basis of the following:

$$(g'(x), \phi(x)) = -(g(x), \phi'(x)) = -\int_0^\infty \phi'(x)dx = \phi(0) = (\delta(x), \phi(x)),$$

as the measure of rational numbers is zero. Therefore,

$$\frac{d^{1.5}}{dx^{1.5}} g(x) = \frac{d^{0.5}}{dx^{0.5}} \delta(x) = \frac{x_+^{-1.5}}{\Gamma(-0.5)} = -\frac{1}{2\sqrt{\pi}} x_+^{-1.5}.$$

We note that the sequential fractional derivative holds in a distribution [26].

For a given function, its classical Riemann–Liouville derivative and/or Caputo derivative [29–31] may not exist in general [32–34]. Even if they do, the Riemann–Liouville derivative and the Caputo derivative are not necessarily the same. However, if $g(x)$ is a distribution in $\mathcal{D}'(R^+)$, then the case is different. Let $m - 1 < \mathrm{Re}\lambda < m \in Z^+$. From Equation (4), we derive that

$$\begin{aligned} g_{-\lambda}(x) &= g(x) * \frac{x_+^{-\lambda-1}}{\Gamma(-\lambda)} = g(x) * \frac{d^m}{dx^m} \frac{x_+^{m-\lambda-1}}{\Gamma(m-\lambda)} \\ &= \frac{d^m}{dx^m} \left(g(x) * \frac{x_+^{m-\lambda-1}}{\Gamma(m-\lambda)} \right) = \frac{x_+^{m-\lambda-1}}{\Gamma(m-\lambda)} * g^{(m)}(x), \end{aligned}$$

which indicates there is no difference between the Riemann–Liouville derivative and the Caputo derivative of the distribution $g(x)$ (both exist clearly). On the basis of this fact, we only call the fractional derivative of the distribution for brevity.

We mention that Podlubny [28] investigated fractional calculus of generalized functions by the distributional convolution and derived many identities of fractional integrals and derivatives related to $\delta(x-a)$, $\theta(x-a)$ and Φ_λ, where a is a constant in R.

We note that the fractional integral (or the Riemann–Liouville fractional integral) $D^{-\alpha} (= D_{0,t}^{-\alpha})$ of order $\alpha \in R^+$ of function $y(t)$ is defined by

$$D^{-\alpha}y(t) = D_{0,t}^{-\alpha}y(t) = \frac{1}{\Gamma(\alpha)}\int_0^t (t-\tau)^{\alpha-1}y(\tau)d\tau \quad (t > 0)$$

if the integral exists.

3. Babenko's Approach with Demonstration

We consider Abel's integral equation of the second kind:

$$y(t) + \frac{\lambda}{\Gamma(\alpha)}\int_0^t (t-\tau)^{\alpha-1}y(\tau)\,d\tau = f(t), \quad t > 0, \tag{7}$$

where we start with $\alpha > 0$ and where λ is a constant. This equation was initially introduced and investigated by Hille and Tamarkin [35] in 1930, who considered Volterra's equation (a more general integral equation):

$$y(t) = f(t) + \lambda\int_0^t K(t,\tau)y(\tau)d\tau$$

by the Liouville–Neumann series. In 1997, Gorenflo and Mainardi studied Abel's integral equations of the first and second kind in their survey paper [36], with emphasis on the method of the Laplace transforms, which is a common treatment of such fractional integral equations. They also outlined the method used by Yu. I. Babenko in his book [37] for solving various types of fractional integral and differential equations. The method itself is close to the Laplace transform method, but it can be used in more cases than the Laplace transform method, such as solving integral equations with variable coefficients. Clearly, it is always necessary to prove convergence of the series obtained as solutions, although it is not a simple task in the general case [28].

In this section, we apply Babenko's method to solve many Abel's integral equations of the second kind, as well as their variants with variable coefficients, and we show convergence of the infinite series by utilizing the rapid growth of the Gamma function. We note that if an infinite series is uniformly convergent in every bounded interval of the variable t, then term-wise integrations and differentiations are allowed.

We can write Equation (7) in the form

$$(1 + \lambda D^{-\alpha})y(t) = f(t)$$

by the Riemann–Liouville fractional integral operator. This implies that

$$y(t) = (1 + \lambda D^{-\alpha})^{-1}f(t) = \sum_{n=0}^{\infty}(-1)^n\lambda^n D^{-\alpha n}f(t) \tag{8}$$

by Babenko's method [28].

Because, for many functions $f(t)$, all the fractional integrals on the right-hand side of Equation (8) can be evaluated, Equation (8) gives the formal solution in the form of a series if it converges.

A two-parameter function of the Mittag–Leffler type is defined by the series expansion

$$E_{\alpha,\beta}(z) = \sum_{k=0}^{\infty}\frac{z^k}{\Gamma(\alpha k + \beta)},$$

where α, $\beta > 0$.

It follows from the Mittag–Leffler function that

$$E_{1,1}(z) = \sum_{k=0}^{\infty} \frac{z^k}{\Gamma(k+1)} = \sum_{k=0}^{\infty} \frac{z^k}{k!} = e^z$$

$$E_{1,2}(z) = \sum_{k=0}^{\infty} \frac{z^k}{\Gamma(k+2)} = \frac{1}{z}\sum_{k=0}^{\infty} \frac{z^{k+1}}{(k+1)!} = \frac{e^z - 1}{z}$$

$$E_{2,1}(z^2) = \sum_{k=0}^{\infty} \frac{z^{2k}}{\Gamma(2k+1)} = \sum_{k=0}^{\infty} \frac{z^{2k}}{(2k)!} = \cosh(z).$$

Demonstration of Examples

Example 1. *Let λ and a be constants and $\alpha > 0$. Then Abel's integral equation:*

$$y(t) + \frac{\lambda}{\Gamma(\alpha)} \int_0^t (t-\tau)^{\alpha-1} y(\tau) d\tau = at^\beta \qquad (9)$$

has the solution

$$y(t) = a\Gamma(\beta+1)t^\beta E_{\alpha,\beta+1}(-\lambda t^\alpha),$$

where $\beta > -1$.

Proof. Clearly, we have

$$D^{-\alpha n} at^\beta = aD^{-\alpha n} t^\beta = \frac{a\Gamma(\beta+1)}{\Gamma(\alpha n + \beta + 1)} t^{\beta + \alpha n}.$$

Hence

$$\begin{aligned}
y(t) &= \sum_{n=0}^{\infty} (-1)^n \lambda^n D^{-\alpha n} at^\beta = \sum_{n=0}^{\infty} (-1)^n \lambda^n \frac{a\Gamma(\beta+1)}{\Gamma(\alpha n + \beta + 1)} t^{\beta + \alpha n} \\
&= a\Gamma(\beta+1)t^\beta \sum_{n=0}^{\infty} (-1)^n \lambda^n \frac{t^{\alpha n}}{\Gamma(\alpha n + \beta + 1)} \\
&= a\Gamma(\beta+1)t^\beta E_{\alpha,\beta+1}(-\lambda t^\alpha),
\end{aligned}$$

which is convergent for all $t \in R^+$. □

This example implies that the following Abel's integral equation:

$$y(t) = 2\sqrt{t} - \int_0^t \frac{y(\tau)}{\sqrt{t-\tau}} d\tau \qquad (10)$$

has the solution

$$y(t) = \sum_{n=1}^{\infty} \frac{(-1)^{n-1}(\pi t)^{n/2}}{\Gamma(n/2+1)}.$$

Indeed, Equation (10) can be converted into

$$y(t) + \frac{\sqrt{\pi}}{\Gamma(1/2)} \int_0^t \frac{y(\tau)}{\sqrt{t-\tau}} d\tau = 2\sqrt{t}.$$

By Equation (9) we come to

$$
\begin{aligned}
y(t) &= 2\Gamma(1/2+1)t^{1/2}E_{1/2,3/2}(-\sqrt{\pi}t^{1/2}) = \sum_{n=0}^{\infty}\frac{(-1)^n(\pi t)^{(n+1)/2}}{\Gamma(n/2+1/2+1)} \\
&= \sum_{n=1}^{\infty}\frac{(-1)^{n-1}(\pi t)^{n/2}}{\Gamma(n/2+1)}
\end{aligned}
$$

using

$$
\Gamma(1/2+1) = \frac{\sqrt{\pi}}{2}.
$$

Example 2. *Abel's integral equation:*

$$
y(t) + \int_0^t \frac{y(\tau)}{\sqrt{t-\tau}}d\tau = t + \frac{4}{3}t^{3/2}
$$

has the solution $y(t) = t$.

Proof. First we note that

$$
\int_0^t \frac{y(\tau)}{\sqrt{t-\tau}}d\tau = \frac{\sqrt{\pi}}{\Gamma(1/2)}\int_0^t \frac{y(\tau)}{\sqrt{t-\tau}}d\tau,
$$

$$
D^{-n/2}t = \frac{t^{(n+2)/2}}{\Gamma(n/2+2)}, \quad \text{and}
$$

$$
D^{-n/2}\frac{4t^{3/2}}{3} = \frac{4}{3}\frac{\Gamma(3/2+1)}{\Gamma(n/2+3/2+1)}t^{(n+3)/2} = \frac{\sqrt{\pi}t^{(n+3)/2}}{\Gamma((n+3)/2+1)}.
$$

We infer from Equation (8) that

$$
\begin{aligned}
y(t) &= \sum_{n=0}^{\infty}(-1)^n\lambda^n D^{-\alpha n}(t+\frac{4}{3}t^{3/2}) = \lim_{m\to\infty}\sum_{n=0}^{m}(-1)^n\pi^{n/2}D^{-n/2}(t+\frac{4}{3}t^{3/2}) \\
&= \lim_{m\to\infty}\Big\{(t+\frac{4}{3}t^{3/2}) - \sqrt{\pi}(\frac{t^{3/2}}{\Gamma(1/2+2)}+\frac{\sqrt{\pi}}{\Gamma(2+1)}t^2)+ \\
&\quad \pi(\frac{t^2}{2}+\frac{4}{3}\frac{t^{3/2+1}}{3/2+1})+\cdots+(-1)^m\pi^{m/2}D^{-m/2}(t+\frac{4}{3}t^{3/2})\Big\} \\
&= t + \lim_{m\to\infty}(-1)^m\pi^{m/2}D^{-m/2}\frac{4}{3}t^{3/2}
\end{aligned}
$$

by cancellations. Furthermore,

$$
\lim_{m\to\infty}(-1)^m\pi^{m/2}D^{-m/2}\frac{4}{3}t^{3/2} = 0
$$

for all $t > 0$. Indeed,

$$
\begin{aligned}
(-1)^m\pi^{m/2}D^{-m/2}\frac{4}{3}t^{3/2} &= \frac{4}{3}(-1)^m\pi^{m/2}\frac{\Gamma(3/2+1)}{\Gamma(m/2+3/2+1)}t^{3/2+m/2} \\
&= \sqrt{\pi}t^{3/2}\frac{(-1)^m(\pi t)^{m/2}}{\Gamma(m/2+3/2+1)},
\end{aligned}
$$

and the series

$$
\sum_{m=0}^{\infty}\frac{(\pi t)^{m/2}}{\Gamma(m/2+3/2+1)} = E_{1/2,3/2+1}(\sqrt{\pi}t)
$$

is convergent. Hence $y(t) = t$ is the solution. □

Remark 1. *Kumar et al. [22] considered Example 2 by applying the aforesaid HPTM and found an approximate solution that converges to the solution $y(t) = t$ for t only between 0 and 1, with more calculations involving Laplace transforms.*

Example 3. *Abel's integral equation:*

$$y(t) + \frac{\lambda}{\Gamma(\alpha)} \int_0^t (t - \tau)^{\alpha-1} y(\tau) \, d\tau = \phi(t)$$

has the solution

$$y(t) = \sum_{n=0}^{\infty} \sum_{k=0}^{\infty} (-1)^n \lambda^n \frac{\phi^{(k)}(0)}{\Gamma(\alpha n + k + 1)} t^{k+\alpha n},$$

where

$$\phi(t) = \sum_{k=0}^{\infty} \frac{\phi^{(k)}(0)}{k!} t^k$$

for t > 0.

Proof. By Equation (8), we obtain

$$\begin{aligned} y(t) &= \sum_{n=0}^{\infty} (-1)^n \lambda^n D^{-\alpha n} \phi(t) \\ &= \sum_{n=0}^{\infty} \sum_{k=0}^{\infty} (-1)^n \lambda^n \frac{\phi^{(k)}(0)}{k!} D^{-\alpha n} t^k \\ &= \sum_{n=0}^{\infty} \sum_{k=0}^{\infty} (-1)^n \lambda^n \frac{\phi^{(k)}(0)}{\Gamma(\alpha n + k + 1)} t^{k+\alpha n}. \end{aligned}$$

It remains to show that the above series is convergent. We note that

$$\lim_{n \to \infty} \frac{\Gamma(n + \alpha)}{\Gamma(n) n^\alpha} = 1$$

implies

$$\Gamma(\alpha n + k + 1) \sim \Gamma(k + 1) n^{\alpha n}$$

when k is large [38]. Therefore, the convergence of

$$\sum_{n=0}^{\infty} \sum_{k=0}^{\infty} (-1)^n \lambda^n \frac{\phi^{(k)}(0)}{\Gamma(\alpha n + k + 1)} t^{k+\alpha n}$$

is equivalent to that of

$$\sum_{k=0}^{\infty} \frac{\phi^{(k)}(0)}{k!} t^k \sum_{n=0}^{\infty} (-1)^n \left(\frac{\lambda t^\alpha}{n^\alpha} \right)^n,$$

which is convergent because it is the product of two convergent series for all $t > 0$.

This clearly implies that the following Abel's integral equation:

$$y(t) + \frac{\lambda}{\Gamma(\alpha)} \int_0^t (t - \tau)^{\alpha-1} y(\tau) \, d\tau = e^t$$

has the solution

$$y(t) = \sum_{n=0}^{\infty} \sum_{k=0}^{\infty} (-1)^n \lambda^n \frac{t^{k+\alpha n}}{\Gamma(\alpha n + k + 1)},$$

and Abel's integral equation:

$$y(t) + \frac{\lambda}{\Gamma(\alpha)} \int_0^t (t - \tau)^{\alpha-1} y(\tau) \, d\tau = \frac{t}{t^2 + b^2}$$

has the solution

$$y(t) = \sum_{n=0}^{\infty} \sum_{k=0}^{\infty} (-1)^n \lambda^n \frac{(-1)^k (2k+1)!}{b^{2k+2} \Gamma(\alpha n + 2k + 2)} t^{\alpha n + 2k + 1}$$

for $0 < t < b$. Indeed,

$$\begin{aligned}
\phi(t) &= \frac{t}{t^2 + b^2} = \frac{t}{b^2 \left(1 + \left(\frac{t}{b}\right)^2\right)} \\
&= \frac{t}{b^2} \sum_{k=0}^{\infty} (-1)^k \left(\frac{t}{b}\right)^{2k} = \sum_{k=0}^{\infty} (-1)^k \frac{t^{2k+1}}{b^{2k+2}},
\end{aligned}$$

which infers that

$$\phi^{(2k+1)}(0) = \frac{(-1)^k (2k+1)!}{b^{2k+2}} \quad \text{and} \quad \phi^{(2k)}(0) = 0.$$

Now we consider some variants of Abel's integral equation. □

Example 4. *Let $\alpha > 0$. Then the integral equation*

$$y(t) + \frac{\lambda t}{\Gamma(\alpha)} \int_0^t (t - \tau)^{\alpha-1} y(\tau) \, d\tau = t$$

has the solution

$$y(t) = t E_{\alpha,2}(-\lambda t^{\alpha+1}).$$

Proof. We can write the original equation in the form

$$(1 + \lambda t D^{-\alpha}) y(t) = t.$$

Therefore

$$\begin{aligned}
y(t) &= (1 + \lambda t D^{-\alpha})^{-1} t = \sum_{n=0}^{\infty} (-1)^n \lambda^n t^n D^{-\alpha n} t \\
&= \sum_{n=0}^{\infty} (-1)^n \lambda^n \frac{t^n}{\Gamma(\alpha n + 2)} t^{\alpha n + 1} \\
&= t \sum_{n=0}^{\infty} (-1)^n \lambda^n \frac{t^{\alpha n + n}}{\Gamma(\alpha n + 2)} = t E_{\alpha,2}(-\lambda t^{\alpha+1}).
\end{aligned}$$

Similarly, the integral equation

$$y(t) + \frac{\lambda t^\beta}{\Gamma(\alpha)} \int_0^t (t - \tau)^{\alpha-1} y(\tau) \, d\tau = t^\gamma \quad \text{for } t > 0$$

has the solution

$$y(t) = \Gamma(\gamma + 1) t^\gamma E_{\alpha,\gamma+1}(-\lambda t^{\alpha+\beta}), \tag{11}$$

where $\gamma > -1$ and $\alpha + \beta > 0$.

Furthermore, the integral equation for $\beta > -1$:

$$y(t) - t^\beta \int_0^t y(\tau)d\tau = 1$$

has the solution $y(t) = e^{t^{\beta+1}}$ by noting that $E_{1,1}(t) = e^t$.

Clearly, the integral equation

$$y(t) - \int_0^t (t - \tau)y(\tau)d\tau = t$$

has the solution $y(t) = tE_{2,2}(t^2) = \sinh(t)$ by noting that $E_{2,2}(t^2) = \sinh(t)/t$, and the integral equation

$$y(t) - \int_0^t (t - \tau)y(\tau)d\tau = 1$$

has the solution $y(t) = \cosh(t)$ as $E_{2,1}(t^2) = \cosh(t)$.

Using the following formula [28]:

$$E_{1,m}(z) = \frac{1}{z^{m-1}} \left\{ e^z - \sum_{k=0}^{m-2} \frac{z^k}{k!} \right\},$$

we derive that the following integral equation for $m = 1, 2, \cdots, \beta > -1$, and $\lambda \neq 0$:

$$y(t) + \lambda t^\beta \int_0^t y(\tau)d\tau = t^m$$

has the solution

$$y(t) = \frac{(-1)^m m!}{\lambda^m t^{\beta m}} \left\{ e^{-\lambda t^{\beta+1}} - \sum_{k=0}^{m-1} \frac{(-\lambda)^k t^{(\beta+1)k}}{k!} \right\}.$$

Indeed, from Equation (11), we infer that the solution is

$$\begin{aligned}
y(t) &= m! t^m E_{1,m+1}(-\lambda t^{\beta+1}) \\
&= \frac{(-1)^m m!}{\lambda^m t^{\beta m}} \left\{ e^{-\lambda t^{\beta+1}} - \sum_{k=0}^{m-1} \frac{(-\lambda)^k t^{(\beta+1)k}}{k!} \right\},
\end{aligned}$$

as $\gamma = m$ and $\alpha = 1$.

Finally, we claim that the following integral equation for $\beta > -1/2$:

$$y(t) + \frac{\lambda t^\beta}{\sqrt{\pi}} \int_0^t \frac{y(\tau)}{\sqrt{t - \tau}} d\tau = 1$$

has the solution $y(t) = E_{1/2,1}(-\lambda t^{\beta+1/2})$, where

$$E_{1/2,1}(z) = \sum_{k=0}^{\infty} \frac{z^k}{\Gamma(k/2 + 1)} = e^{z^2} \text{erfc}(-z)$$

and $\text{erfc}(z)$ is the error function complement defined by

$$\text{erfc}(z) = \frac{2}{\sqrt{\pi}} \int_z^\infty e^{-t^2} dt.$$

To end this section, we point out that some exact solutions of equations considered in the examples can be easily obtained using the Laplace transform mentioned earlier; for example, applying the Laplace transform to Equation (9) from Example 1, we obtain

$$y^*(s) = \mathcal{L}\{y(t)\} = a\Gamma(\beta + 1)\frac{s^{\alpha - (\beta + 1)}}{s^\alpha + \lambda}$$

by the formula

$$\int_0^\infty t^\beta e^{-st} dt = \frac{\Gamma(\beta + 1)}{s^{\beta + 1}}, \quad \operatorname{Res} > 0.$$

This implies the following by the inverse Laplace transform and Equation (1.80) in [28]:

$$y(t) = a\Gamma(\beta + 1)t^\beta E_{\alpha, \beta + 1}(-\lambda t^\alpha).$$

Similarly, for the equation from Example 2, we immediately derive

$$y^*(s) = \mathcal{L}\{y(t)\} = \frac{1}{s^2}$$

from

$$\mathcal{L}\{t + \frac{4}{3}t^{3/2}\} = \frac{1}{s^2} + \frac{\sqrt{\pi}}{s^{2.5}},$$

which indicates that $y(t) = t$. However, it seems much harder for the Laplace transform to deal with the variants of Abel's integral equation with variable coefficients, such as

$$y(t) + \lambda t^\beta \int_0^t y(\tau) d\tau = t^m,$$

which is solved above by Babenko's approach. □

4. Abel's Integral Equation in Distributions

We let $f(t) \in \mathcal{D}'(R^+)$ be given. We now study Abel's integral equation of the second kind:

$$y(t) + \frac{\lambda}{\Gamma(\alpha)} \int_0^t (t - \tau)^{\alpha - 1} y(\tau) \, d\tau = f(t) \tag{12}$$

in the distributional space $\mathcal{D}'(R^+)$, where $\alpha \in R$ and λ is a constant.

Clearly, Equation (12) is equivalent to the convolutional equation:

$$(\delta + \lambda \Phi_\alpha) * y(t) = f(t)$$

in $\mathcal{D}'(R^+)$, although it is undefined in the classical sense for $\alpha \leq 0$. We note that the distributional convolution $\Phi_\alpha * y(t)$ is well defined for arbitrary $\alpha \in R$ if $y(t) \in \mathcal{D}'(R^+)$ by Definition 1 (b).

Applying Babenko's method, we can write out $y(t)$ as

$$y(t) = (\delta + \lambda \Phi_\alpha)^{-1} * f(x),$$

where $(\delta + \lambda \Phi_\alpha)^{-1}$ is the distributional inverse operator of $\delta + \lambda \Phi_\alpha$ in terms of convolution. Using the binomial expression of $(\delta + \lambda \Phi_\alpha)^{-1}$, we can formally obtain

$$y(t) = \sum_{n=0}^\infty (-1)^n \lambda^n \Phi_\alpha^n * f(t) = \sum_{n=0}^\infty (-1)^n \lambda^n \Phi_{n\alpha} * f(t) \tag{13}$$

by using the following formula [25]:

$$\Phi_\alpha^n = \underbrace{\Phi_\alpha * \Phi_\alpha * \cdots * \Phi_\alpha}_{n \text{ times}} = \Phi_{n\alpha},$$

and δ is a unit distribution in terms of convolution.

Because, for many distributions $f(t)$, all the fractional integrals (if $\alpha > 0$) or derivatives (if $\alpha < 0$) on the right-hand side of Equation (13) can be evaluated, Equation (13) gives the formal solution in the form of a series if it converges. We must point out that Equation (12) becomes the differential equation:

$$y + \lambda y * \delta^{(m)} = y + \lambda y^{(m)} = f$$

if $\alpha = -m$ (undefined in the classical sense) because of Equation (1), which can be converted into a system of linear differential equations for consideration [24].

Example 5. *Abel's integral equation:*

$$y(t) + \int_0^t \frac{y(\tau)}{\sqrt{t - \tau}} \, d\tau = t_+^{-1.5} \tag{14}$$

has its solution in the space $\mathcal{D}'(R^+)$:

$$y(t) = t_+^{-1.5} + 2\pi\delta(t) - 2\pi t_+^{-1/2} + 2\pi^2 E_{1/2,1}(-\sqrt{\pi t_+}).$$

Proof. Equation (14) can be written as

$$y(t) + \frac{\sqrt{\pi}}{\Gamma(1/2)} \int_0^t \frac{y(\tau)}{\sqrt{t - \tau}} \, d\tau = t_+^{-1.5}.$$

From Equation (13),

$$
\begin{aligned}
y(t) &= \sum_{n=0}^{\infty} (-1)^n \pi^{n/2} \Phi_{1/2}^n * t_+^{-1.5} = \sum_{n=0}^{\infty} (-1)^n \pi^{n/2} \Phi_{n/2} * \frac{t_+^{-1.5}}{\Gamma(-0.5)} \Gamma(-0.5) \\
&= \sum_{n=0}^{\infty} (-1)^n \pi^{n/2} \Gamma(-0.5) \Phi_{n/2} * \Phi_{-0.5} = \sum_{n=0}^{\infty} (-1)^n \pi^{n/2} \Gamma(-0.5) \Phi_{n/2-0.5} \\
&= \sum_{n=0}^{\infty} (-1)^n \pi^{n/2} \Gamma(-0.5) \frac{t_+^{n/2-1.5}}{\Gamma(n/2 - 0.5)} \\
&= t_+^{-1.5} + 2\pi\delta(t) - 2\pi t_+^{-1/2} + 2 \sum_{n=3}^{\infty} (-1)^{n+1} \pi^{(n+1)/2} \frac{t_+^{n/2-1.5}}{\Gamma(n/2 - 0.5)} \\
&= t_+^{-1.5} + 2\pi\delta(t) - 2\pi t_+^{-1/2} + 2\pi^2 E_{1/2,1}(-\sqrt{\pi t_+})
\end{aligned}
$$

using $\Gamma(-0.5) = -2\sqrt{\pi}$. $\quad\square$

Remark 2. *Equation (14) cannot be discussed in the classical sense, because the fractional integral of $t_+^{-1.5}$ does not exist in the normal sense. Clearly, the distribution $t_+^{-1.5} + 2\pi\delta(t)$ is a singular generalized function in $\mathcal{D}'(R^+)$, while $-2\pi t_+^{-1/2} + 2\pi^2 E_{1/2,1}(-\sqrt{\pi t_+})$ is regular (locally integrable).*

Example 6. *For $m = 0, 1, 2, \cdots$, the integral equation*

$$y(t) + t^m \int_0^t \frac{y(\tau)}{\sqrt{t - \tau}} \, d\tau = \delta(t)$$

has the solution

$$y(t) = \delta(t) - t_+^{m-1/2} + \pi\, t_+^{2m} E_{1/2,1}(-\sqrt{\pi}t_+^{m+1/2}).$$

Proof. Clearly,

$$
\begin{aligned}
y(t) &= \sum_{n=0}^{\infty} (-1)^n \pi^{n/2} t^{mn} \Phi_{n/2} * \delta(t) = \sum_{n=0}^{\infty} (-1)^n \pi^{n/2} t^{mn} \Phi_{n/2} \\
&= \delta(t) - \pi^{1/2} t^m \frac{t_+^{-1/2}}{\Gamma(1/2)} + \sum_{n=2}^{\infty} (-1)^n \pi^{n/2} t^{mn} \Phi_{n/2} \\
&= \delta(t) - t_+^{m-1/2} + \sum_{n=2}^{\infty} (-1)^n \pi^{n/2} \frac{t_+^{mn-1+n/2}}{\Gamma(n/2)} \\
&= \delta(t) - t_+^{m-1/2} + \pi\, t_+^{2m} \sum_{k=0}^{\infty} (-1)^k \pi^{k/2} \frac{t_+^{mk+k/2}}{\Gamma(k/2+1)} \\
&= \delta(t) - t_+^{m-1/2} + \pi\, t_+^{2m} E_{1/2,1}(-\sqrt{\pi}t_+^{m+1/2}),
\end{aligned}
$$

which is a distribution in $\mathcal{D}'(R^+)$.

We note that if f is a distribution in $\mathcal{D}'(R)$ and g is an infinitely differentiable function, then the product fg is defined by

$$(fg, \phi) = (f, g\phi)$$

for all functions $\phi \in \mathcal{D}(R)$. Therefore, the product

$$t^m \int_0^t \frac{y(\tau)}{\sqrt{t-\tau}}\, d\tau$$

is well defined, as t^m is an infinitely differentiable function.

In particular, Abel's integral equation:

$$y(t) + \int_0^t \frac{y(\tau)}{\sqrt{t-\tau}}\, d\tau = \delta(t)$$

has the solution

$$y(t) = \delta(t) - t_+^{-1/2} + \pi\, E_{1/2,1}(-\sqrt{\pi}t_+),$$

and the integral equation

$$y(t) + t \int_0^t \frac{y(\tau)}{\sqrt{t-\tau}}\, d\tau = \delta(t)$$

has the solution

$$y(t) = \delta(t) - t_+^{1/2} + \pi\, t_+^2 E_{1/2,1}(-\sqrt{\pi}t_+^{1+1/2}).$$

We mention that Abel's integral equations or integral equations with more general weakly singular kernels play important roles in solving partial differential equations, in particular, parabolic equations in which naturally the independent variable has the meaning of time. Gorenflo and Mainardi [36] described the occurrence of Abel's integral equations of the first and second kind in the problem of the heating (or cooling) of a semi-infinite rod by influx (or efflux) of heat across the boundary into (or from) its interior. They used Abel's integral equations to solve the following equation of heat flow:

$$u_t - u_{xx} = 0, \quad u = u(x,t)$$

in the semi-infinite intervals $0 < x < \infty$ and $0 < t < \infty$ of space and time, respectively. In this dimensionless equation, $u = u(x,t)$ refers to temperature. Assuming a vanishing initial temperature, that is, $u(x,0) = 0$ for $0 < x < \infty$, and a given influx across the boundary $x = 0$ from $x < 0$ to $x > 0$,

$$-u_x(0,t) = p(t).$$

The interested readers are referred to [36] for the detailed methods and further references. □

5. Conclusions

Applying Babenko's method and fractional calculus, we have studied Abel's integral equation of the second kind as well as its variants with variable coefficients, and we have solved many types of different integral equations by showing the convergence of infinite series using the rapid growth of the Gamma function. In particular, we have discussed Abel's equation in the distributional sense for any arbitrary $\alpha \in R$ by fractional operations of generalized functions for the first time, and we have derived several new and interesting results that cannot be realized in the classical sense or by the Laplace transform. For example, Abel's integral equation:

$$y(t) + \int_0^t \frac{y(\tau)}{\sqrt{t-\tau}}\, d\tau = t_+^{-\sqrt{2}} + \delta^{(m)}(t)$$

cannot be considered in the classical sense because of the term $t_+^{-\sqrt{2}} + \delta^{(m)}(t)$, but it is well defined and solvable in the distributional sense.

Acknowledgments: This work is partially supported by the Natural Sciences and Engineering Research Council of Canada (2017-00001) and Brandon University Research Grant. The authors are grateful to the reviewers for the careful reading of the paper with several productive suggestions and corrections, which certainly improved its quality.

Author Contributions: The order of the author list reflects contributions to the paper.

Conflicts of Interest: The authors declare no conflict of interest.

References

1. Mann, W.R.; Wolf, F. Heat transfer between solids and gases under nonlinear boundary conditions. *Q. Appl. Math.* **1951**, *9*, 163–184
2. Goncerzewicz, J.; Marcinkowska, H.; Okrasinski, W.; Tabisz, K. On percolation of water from a cylindrical reservoir into the surrounding soil. *Appl. Math.* **1978**, *16*, 249–261.
3. Keller, J.J. Propagation of simple nonlinear waves in gas filled tubes with friction. *Z. Angew. Math. Phys.* **1981**, *32*, 170–181.
4. Wang, J.R.; Zhu, C.; Fečkan, M. Analysis of Abel-type nonlinear integral equations with weakly singular kernels. *Bound. Value Probl.* **2014**, *2014*, 20.
5. Atkinson, K.E. An existence theorem for Abel integral equations. *SIAM J. Math. Anal.* **1974**, *5*, 729–736.
6. Bushell, P.J.; Okrasinski, W. Nonlinear Volterra integral equations with convolution kernel. *J. Lond. Math. Soc.* **1990**, *41*, 503–510.
7. Gorenflo, R.; Vessella, S. *Abel Integral Equations: Analysis and Applications*; Springer: Berlin, Germany, 1991.
8. Okrasinski, W. Nontrivial solutions to nonlinear Volterra integral equations. *SIAM J. Math. Anal.* **1991**, *22*, 1007–1015.
9. Gripenberg, G. On the uniqueness of solutions of Volterra equations. *J. Integral Equ. Appl.* **1990**, *2*, 421–430.
10. Mydlarczyk, W. The existence of nontrivial solutions of Volterra equations. *Math. Scand.* 1991, *68*, 83–88.
11. Kilbas, A.A.; Saigo, M. On solution of nonlinear Abel-Volterra integral equation. *J. Math. Anal. Appl.* **1999**, *229*, 41–60.
12. Karapetyants, N.K.; Kilbas, A.A.; Saigo, M. Upper and lower bounds for solutions of nonlinear Volterra convolution integral equations with power nonlinearity. *J. Integral Equ. Appl.* **2000**, *12*, 421–448.
13. Lima, P.; Diogo, T. Numerical solution of nonuniquely solvable Volterra integral equation using extrapolation methods. *J. Comput. App. Math.* **2002**, *140*, 537–557.
14. Tamarkin, J.D. On integrable solutions of Abel's integral equation. *Ann. Math.* **1930**, *31*, 219–229.
15. Sumner, D.B. Abel's integral equation as a convolution of transform. *Proc. Am. Math. Soc.* **1956**, *7*, 82–86.

16. Minerbo, G.N.; Levy, M.E. Inversion of Abel's integral equation by means of orthogonal polynomials. *SIAM J. Numer. Anal.* **1969**, *6*, 598–616.
17. Hatcher, J.R. A nonlinear boundary problem. *Proc. Am. Math. Soc.* **1985**, *95*, 441–448.
18. Singh, O.P.; Singh, V.K.; Pandey, R.K. A stable numerical inversion of Abel's integral equation using almost Bernstein operational matrix. *J. Quant. Spectrosc. Radiat. Transf.* **2010**, *111*, 245–252.
19. Li, M.; Zhao, W. Solving Abel's type integral equation with Mikusinski's operator of fractional order. *Adv. Math. Phys.* **2013**, *2013*, doi:10.1155/2013/806984.
20. Jahanshahi, S.; Babolian, E.; Torres, D.; Vahidi, A. Solving Abel integral equations of first kind via fractional calculus. *J. King Saud Univ. Sci.* **2015**, *27*, 161–167.
21. Saleh, M.H.; Amer, S.M.; Mohamed, D.S.; Mahdy, A.E. Fractional calculus for solving generalized Abel's integral equations suing Chebyshev polynomials. *Int. J. Comput. Appl.* **2014**, *100*, 19–23.
22. Kumar, S.; Kumar, A.; Kumar, D.; Singh, J.; Singh, A. Analytical solution of Abel integral equation arising in astrophysics via Laplace transform. *J. Egypt. Math. Soc.* **2015**, *23*, 102–107.
23. Li, C.; Li, C.P.; Kacsmar, B.; Lacroix, R.; Tilbury, K. The Abel integral equations in distributions. *Adv. Anal.* **2017**, *2*, 88–104.
24. Gel'fand, I.M.; Shilov, G.E. *Generalized Functions*; Academic Press: New York, NY, USA, 1964; Volume I.
25. Li, C. Several results of fractional derivatives in $\mathcal{D}'(R^+)$. *Fract. Calc. Appl. Anal.* **2015**, *18*, 192–207.
26. Matignon, D. Stability results for fractional differential equations with applications to control processing. In Proceedings of the IEEE-SMC Computational Engineering in System Applications, Lille, France, 10 February 1996; Volume 2, pp. 963–968.
27. Mainardi, F. Fractional relaxation-oscillation and fractional diffusion-wave phenomena. *Chaos Solitons Fractals* **1996**, *7*, 1461–1477.
28. Podlubny, I. *Fractional Differential Equations*; Academic Press: New York, NY, USA, 1999.
29. Li, C.P.; Qian, D.; Chen, Y. On Riemann-Liouville and Caputo derivatives. *Discret. Dyn. Nat. Soc.* **2011**, doi:10.1155/2011/562494.
30. Li, C.P.; Zhao, Z. Introduction to fractional integrability and differentiability. *Eur. Phys. J. Spec. Top.* **2011**, *193*, 5–26.
31. Li, C.P.; Zhang, F.; Kurths, J.; Zeng, F. Equivalent system for a multiple-rational-order fractional differential system. *Phil. Trans. R. Soc. A* **2013**, *371*, 20120156.
32. Li, C.P.; Deng, W. Remarks on fractional derivatives. *Appl. Math. Comput.* **2007**, *187*, 777–784.
33. Kilbas, A.A.; Srivastava, H.M.; Trujillo, J.J. *Theory and Applications of Fractional Differential Equations*; Elsevier: New York, NY, USA, 2006.
34. Li, C.; Li, C.P. On defining the distributions $(\delta)^k$ and $(\delta')^k$ by fractional derivatives. *Appl. Math. and Comput.* **2014**, *246*, 502–513.
35. Hille, E.; Tamarkin, J.D. On the theory of linear integral equations. *Ann. Math.* **1930**, *31*, 479–528.
36. Gorenflo, R.; Mainardi, F. Fractional Calculus: Integral and Differential Equations of Fractional Order. In *Fractals and Fractional Calculus in Continuum Mechanics*; Springer: New York, NY, USA, 1997; pp. 223–276.
37. Babenko, Y.I. *Heat and Mass Transfer*; Khimiya: Leningrad, Russian, 1986. (In Russian)
38. Davis, P.J. Leonhard Euler's Integral: A Historical Profile of the Gamma Function. *Am. Math. Mon.* **1959**, *66*, 849–869.

mathematics

MDPI

Article

Several Results of Fractional Differential and Integral Equations in Distribution

Chenkuan Li [1,*], Changpin Li [2] and Kyle Clarkson [1]

[1] Department of Mathematics and Computer Science, Brandon University, Brandon, MB R7A 6A9, Canada; kyleclarkson17@hotmail.com

[2] Department of Mathematics, Shanghai University, Shanghai 200444, China; lcp@shu.edu.cn

* Correspondence: lic@brandonu.ca

Received: 10 May 2018; Accepted: 6 June 2018; Published: 8 June 2018

Abstract: This paper is to study certain types of fractional differential and integral equations, such as $\theta(x - x_0)g(x) = \frac{1}{\Gamma(\alpha)} \int_0^x (x - \zeta)^{\alpha-1} f(\zeta)d\zeta$, $y(x) + \int_0^x \frac{y(\tau)}{\sqrt{x-\tau}} \, d\tau = x_+^{-\sqrt{2}} + \delta(x)$, and $x_+^k \int_0^x y(\tau)(x - \tau)^{\alpha-1} \, d\tau = \delta^{(m)}(x)$ in the distributional sense by Babenko's approach and fractional calculus. Applying convolutions and products of distributions in the Schwartz sense, we obtain generalized solutions for integral and differential equations of fractional order by using the Mittag-Leffler function, which cannot be achieved in the classical sense including numerical analysis methods, or by the Laplace transform.

Keywords: distribution; fractional calculus; convolution; Abel's integral equation; product; Mittag-Leffler function

MSC: 46F10; 26A33; 45E10; 45G05

1. Introduction

Fractional calculus deals with integrals and derivatives of arbitrary order, which unifies and extends integer-order differentiation and n-fold integration. Up to now, fractional operators have been applied in various areas, such as anomalous diffusion, long-range interactions, long-memory processes and materials, waves in liquids, and physics. Recently, Zimbardo et al. [1] generalized the Parker equation, which describes the acceleration and transport of energetic particles in astrophysical plasmas, to the case of anomalous by Caputo fractional derivatives. As far as we know, fractional calculus is one of best tools to construct certain electro-chemical problems and characterizes long-term behaviors, allometric scaling laws, nonlinear operations of distributions [2] and so on.

An integral equation is an equation containing an unknown function under an integral sign. Integral equations are useful and powerful mathematical tools in both pure and applied mathematics. They have various applications in numerous physical problems, chemistry, biology, electronics and mechanics [3–5]. Many initial and boundary value problems associated with ordinary (or partial) differential equations can be transformed into problems of solving integral equations [6]. For example, Gorenflo and Mainardi [7] provided interesting applications of Abel's integral equations of the first and second kind in solving the partial differential equation which describes the problem of the heating (or cooling) of a semi-infinite rod by influx (or efflux) of heat across the boundary into (or from) its interior. There have been lots of approaches, including numerical analysis, thus far to studying fractional differential and integral equations, including Abel's equations, with many applications [8–21]. Recently, Li et al. [22,23] studied integral equations associated with Abel's types in the distributional (Schwartz) sense, based on new fractional calculus of distributions and derived fresh results which are not achievable in the classical sense. On the other hand, the development of science has led to formation of many physical and engineering problems that can be mathematically represented by differential equations. For instance, problems from electric circuits, chemical kinetics, and transfer of heat can all be characterized as differential equations [24].

We briefly introduce the necessary concepts and definitions of fractional calculus of distributions in $\mathcal{D}'(R^+)$ in Section 2, where we demonstrate several examples of computing fractional derivatives as well as applications to solving Abel's integral equations of the first kind for arbitrary $\alpha \in R$. In Section 3, we present Babenko's approach to solving several fractional differential and integral equations based on new convolution and product of generalized functions (an active area in distribution theory). We often obtain an infinite series, related to the Mittag-Leffler function, as the solution of integral or differential equation. We imply a number of novel results, which cannot be derived or approximated by numerical analysis methods or by the Laplace transform, since solutions are in the distributional sense in general.

2. Fractional Calculus in $\mathcal{D}'(R^+)$

In order to investigate fractional integral and differential equations in the generalized sense, we briefly introduce the following basic concepts, with several interesting examples of solving Abel's integral equations in distribution. Let $\mathcal{D}(R)$ be the Schwartz space (testing function space) [25] of infinitely differentiable functions with compact support in R, and $\mathcal{D}'(R)$ the (dual) space of distributions defined on $\mathcal{D}(R)$. A sequence $\phi_1, \phi_2, \cdots, \phi_n, \cdots$ goes to zero in $\mathcal{D}(R)$, if and only if these functions vanish outside a certain fixed bounded set, and converge to zero uniformly together with their derivatives of any order. Clearly, $\mathcal{D}(R)$ is not empty since it contains the following function

$$\phi(x) = \begin{cases} e^{-\frac{1}{1-x^2}} & \text{if } |x| < 1, \\ 0 & \text{otherwise.} \end{cases}$$

Evidently, any locally integrable function $f(x)$ on R is a (regular) distribution in $\mathcal{D}'(R)$ as

$$(f(x), \phi(x)) = \int_{-\infty}^{\infty} f(x)\phi(x)dx$$

is well defined. Hence f is linear and continuous on $\mathcal{D}(R)$. Furthermore, the functional $\delta(x - x_0)$ on $\mathcal{D}(R)$ given by

$$(\delta(x - x_0), \phi(x)) = \phi(x_0)$$

is a member of $\mathcal{D}'(R)$, according to the topological structure of the Schwartz testing function space.

Let $f \in \mathcal{D}'(R)$. The distributional derivative f' (or df/dx), is defined as

$$(f', \phi) = -(f, \phi')$$

for $\phi \in \mathcal{D}(R)$. Therefore,

$$(\delta^{(n)}(x - x_0), \phi(x)) = (-1)^n(\delta(x - x_0), \phi^{(n)}(x)) = (-1)^n \phi^{(n)}(x_0).$$

Let

$$g(x) = x_+ = \begin{cases} x & \text{if } x > 0, \\ 0 & \text{otherwise.} \end{cases}$$

As an example, we will find the distributional derivative of g (note that this function is not differentiable at $x = 0$ in the classical sense). Indeed, using integration by parts, we derive

$$(g'(x), \phi(x)) = -(g(x), \phi'(x)) = -\int_0^\infty x\phi'(x)dx = \int_0^\infty \phi(x)dx = (\theta(x), \phi(x))$$

which infers that

$$g'(x) = \theta(x)$$

where $\theta(x)$ is the Heaviside function.

Let $\text{Re}\lambda > -n - 1, \lambda \neq -1, -2, \cdots, -n$. Then the distribution x_+^λ [25] is defined by

$$(x_+^\lambda, \phi(x)) = \int_0^\infty x^\lambda \phi(x) dx = \int_0^1 x^\lambda [\phi(x) - \phi(0) - \cdots - \frac{t^{n-1}}{(n-1)!}\phi^{(n-1)}(0)]dt$$

$$+ \int_1^\infty x^\lambda \phi(x) dx + \sum_{k=1}^n \frac{\phi^{(k-1)}(0)}{(k-1)!(\lambda+k)}.$$

Clearly, the right-hand side regularizes the integral on the left. This defines the distribution x_+^λ for $\text{Re}\lambda \neq -1, -2, \ldots$.

Assume that f and g are distributions in $\mathcal{D}'(R^+)$. Then the convolution $f * g$ is well defined by the equation [25]

$$((f * g)(x), \phi(x)) = (g(x), (f(y), \phi(x + y)))$$

for $\phi \in \mathcal{D}(R)$. This also implies that

$$f * g = g * f \quad \text{and} \quad (f * g)' = g' * f = g * f'.$$

Let $\mathcal{D}'(R^+)$ be the subspace of $\mathcal{D}'(R)$ with support contained in R^+. It follows from [25–27] that $\Phi_\lambda = \frac{x_+^{\lambda-1}}{\Gamma(\lambda)} \in \mathcal{D}'(R^+)$ is an entire analytic function of λ on the complex plane, and

$$\frac{x_+^{\lambda-1}}{\Gamma(\lambda)}\bigg|_{\lambda=-n} = \delta^{(n)}(x), \quad \text{for} \quad n = 0, 1, 2, \ldots \tag{1}$$

which plays an important role in solving fractional differential equations by using the distributional convolutions. Let λ and μ be arbitrary numbers, then the following identity

$$\Phi_\lambda * \Phi_\mu = \Phi_{\lambda+\mu} \tag{2}$$

is satisfied [23].

Let λ be an arbitrary complex number and $g(x)$ be a distribution in $\mathcal{D}'(R^+)$. We define the primitive of order λ of g as the distributional convolution

$$g_\lambda(x) = g(x) * \frac{x_+^{\lambda-1}}{\Gamma(\lambda)} = g(x) * \Phi_\lambda. \tag{3}$$

Note that this is well defined since the distributions g and Φ_λ are in $\mathcal{D}'(R^+)$. We shall write the convolution

$$g_{-\lambda} = \frac{d^\lambda}{dx^\lambda} g = g(x) * \Phi_{-\lambda}$$

as the fractional derivative of the distribution g of order λ if $\text{Re}\lambda \geq 0$, and $\frac{d^\lambda}{dx^\lambda} g$ is interpreted as the fractional integral if $\text{Re}\lambda < 0$.

We should add that Gorenflo and Mainardi [7] formally presented the derivative of order n (non-negative integer) of g by the generalized convolution between Φ_{-n} and g,

$$\frac{d^n}{dx^n} g(x) = g^{(n)}(x) = \Phi_{-n}(x) * g(x) = \int_0^x g(\tau)\delta^{(n)}(x - \tau)d\tau.$$

based on the well known properties

$$\int_{-\infty}^{\infty} g(\tau)\delta^{(n)}(\tau - x)d\tau = (-1)^n g^{(n)}(x), \quad \delta^{(n)}(x - \tau) = (-1)^n \delta^{(n)}(\tau - x).$$

Then, a formal definition of the fractional derivative of order λ could be

$$\Phi_{-\lambda}(x) * g(x) = \frac{1}{\Gamma(-\lambda)} \int_0^x \frac{g(\tau)}{(x - \tau)^{1+\lambda}} d\tau, \quad \lambda \in R^+.$$

Note that this convolution is in the distributional sense (and it does not exist in the classical sense as the kernel $\Phi_{-\lambda}(x)$ is not locally integrable).

Example 1. *Let*

$$g(x) = \begin{cases} 1 & \text{if } 0 < x < 1 \text{ and } x \text{ is irrational,} \\ 0 & \text{otherwise.} \end{cases}$$

Then,

$$g^{(1.5)}(x) = \delta^{(0.5)}(x) - \delta^{(0.5)}(x - 1)$$

where

$$\delta^{(0.5)}(x) = -\frac{x_+^{-1.5}}{2\sqrt{\pi}} \quad \text{and} \quad \delta^{(0.5)}(x - 1) = -\frac{(x-1)_+^{-1.5}}{2\sqrt{\pi}}.$$

In fact, we have for $\phi \in \mathcal{D}(R)$

$$\begin{aligned} (g'(x), \phi(x)) &= -(g(x), \phi'(x)) = -\int_{-\infty}^{\infty} g(x)\phi'(x)dx = -\int_0^1 \phi'(x)dx \\ &= -\phi(1) + \phi(0) = (\delta(x) - \delta(x - 1), \phi(x)) \end{aligned}$$

as the measure of rational numbers is zero. This indicates that

$$g'(x) = \delta(x) - \delta(x - 1).$$

Obviously, by the sequential fractional derivative law, we come to

$$g^{(1.5)}(x) = \frac{d^{0.5}}{dx^{0.5}} g'(x) = \delta^{(0.5)}(x) - \delta^{(0.5)}(x - 1)$$

and

$$\delta^{(0.5)}(x) = \frac{d^{0.5}}{dx^{0.5}}\delta(x) = \frac{d^{0.5}}{dx^{0.5}}\Phi_0 = \frac{x_+^{-1.5}}{\Gamma(-0.5)} = -\frac{x_+^{-1.5}}{2\sqrt{\pi}}$$

using

$$\Gamma(-0.5) = -2\sqrt{\pi}.$$

Furthermore,

$$g^{(0.5)}(x) = \frac{d^{-0.5}}{dx^{-0.5}} g'(x) = \delta^{(-0.5)}(x) - \delta^{(-0.5)}(x - 1)$$

where

$$\delta^{(-0.5)}(x) = \Phi_{0.5} * \Phi_0 = \Phi_{0.5} = \frac{x_+^{-0.5}}{\Gamma(0.5)} = \frac{x_+^{-0.5}}{\sqrt{\pi}}$$

$$\delta^{(-0.5)}(x - 1) = \frac{(x-1)_+^{-0.5}}{\sqrt{\pi}}$$

which are regular distributions (locally integrable functions on R). In general, we can get

$$g(x) = \theta(x) - \theta(x-1)$$

distributionally by noting that

$$\theta(x) = \Phi_1(x).$$

Then, we imply that for any complex number $\alpha \in C$

$$
\begin{aligned}
g^{(\alpha)}(x) &= \theta^{(\alpha)}(x) - \theta^{(\alpha)}(x-1) = \Phi_{-\alpha}(x) * \Phi_1(x) + \Phi_{-\alpha}(x) * \Phi_1(x-1) \\
&= \Phi_{1-\alpha}(x) + \Phi_{1-\alpha}(x-1) = \frac{x_+^{-\alpha}}{\Gamma(1-\alpha)} + \frac{(x-1)_+^{-\alpha}}{\Gamma(1-\alpha)}.
\end{aligned}
$$

In particular,

$$g^{(5/2)} = \frac{3x_+^{-5/2}}{4\sqrt{\pi}} + \frac{3(x-1)_+^{-5/2}}{4\sqrt{\pi}}$$

by

$$\Gamma(-3/2) = \frac{4}{3}\sqrt{\pi}.$$

   Now we are ready to present the following theorem.

Theorem 1. *Let* $g(x)$ *be an infinitely differentiable function on* $[0, \infty]$ *and* f *be an unknown distribution in* $\mathcal{D}'(R^+)$. *Then the generalized Abel's integral equation of the first kind*

$$\theta(x - x_0)g(x) = \frac{1}{\Gamma(\alpha)} \int_0^x (x-\zeta)^{\alpha-1} f(\zeta)d\zeta \tag{4}$$

has the solution

$$f(x) = \theta(x - x_0)g(x) * \Phi_{-\alpha}$$

where α *is any real number in R and* $x_0 \geq 0$. *In particular, we have four different cases depending on the value of* α.

(i)   *If* $m < \alpha < m+1$ *for* $m = 0, 1, \ldots$, *then*

$$
\begin{aligned}
f(x) &= \frac{d^{m+1}}{dx^{m+1}} \theta(x-x_0)g(x) * \frac{x_+^{-\alpha+m}}{\Gamma(-\alpha+m+1)} \\
&= g(x_0)\frac{(x-x_0)_+^{-\alpha}}{\Gamma(-\alpha+1)} + \cdots + g^{(m)}(x_0)\frac{(x-x_0)_+^{-\alpha+m}}{\Gamma(-\alpha+m+1)} \\
&\quad + \frac{1}{\Gamma(-\alpha+m+1)} \int_{x_0}^x g^{(m+1)}(\zeta)(x-\zeta)^{-\alpha+m}d\zeta
\end{aligned}
$$

   for $x \geq x_0$.

(ii)   *If* $\alpha = 1, 2, \ldots$, *then*

$$f(x) = g(x_0)\delta^{(\alpha-1)}(x-x_0) + \cdots + g^{(\alpha-1)}(x_0)\delta(x-x_0) + \theta(x-x_0)g^{(\alpha)}(x).$$

(iii)   *If* $\alpha = 0$, *then* $f(x) = \theta(x-x_0)g(x)$.

(iv)   *If* $\alpha < 0$, *then for* $x \geq x_0$

$$f(x) = \frac{1}{\Gamma(-\alpha)} \int_{x_0}^x g(\zeta)(x-\zeta)^{-\alpha-1}d\zeta$$

   which is well defined.

Proof. Clearly, we can convert Equation (4) into

$$\theta(x - x_0)g(x) = f(x) * \Phi_\alpha$$

which infers that

$$\theta(x - x_0)g(x) * \Phi_{-\alpha} = (f(x) * \Phi_\alpha) * \Phi_{-\alpha} = f(x) * \Phi_0 = f(x) * \delta(x) = f(x)$$

by using Equations (1) and (2). Assuming $m < \alpha < m + 1$ for $m = 0, 1, \ldots$, we derive that

$$\Phi_{-\alpha} = \frac{d^{m+1}}{dx^{m+1}} \frac{x_+^{-\alpha+m}}{\Gamma(-\alpha + m + 1)}$$

where

$$\frac{x_+^{-\alpha+m}}{\Gamma(-\alpha + m + 1)}$$

is a locally integrable function on R since $-1 < -\alpha + m < 0$. Hence,

$$f(x) = \theta(x - x_0)g(x) * \frac{d^{m+1}}{dx^{m+1}} \frac{x_+^{-\alpha+m}}{\Gamma(-\alpha + m + 1)} = \frac{d^{m+1}}{dx^{m+1}} \theta(x - x_0)g(x) * \frac{x_+^{-\alpha+m}}{\Gamma(-\alpha + m + 1)}.$$

In particular, choosing $m = 0$ (then $0 < \alpha < 1$) and $x_0 = 0$ we come to

$$f(x) = \frac{1}{\Gamma(1 - \alpha)} \int_0^x \frac{g'(\zeta)}{(x - \zeta)^\alpha} d\zeta$$

for a differentiable function $g(x)$ (in the classical sense). This is the solution for the classical Abel's integral equation.

Obviously, we get from integration by parts,

$$(\frac{d}{dx}\theta(x - x_0)g(x), \phi(x)) = -\int_{x_0}^\infty g(x)\phi'(x)dx = g(x_0)\phi(x_0) + \int_{x_0}^\infty \phi(x)g'(x)dx$$

where ϕ is the testing function. This implies that

$$\frac{d}{dx}\theta(x - x_0)g(x) = g(x_0)\delta(x - x_0) + \theta(x - x_0)g'(x).$$

By mathematical induction,

$$\begin{aligned}
\frac{d^{m+1}}{dx^{m+1}}\theta(x - x_0)g(x) &= g(x_0)\delta^{(m)}(x - x_0) + g'(x_0)\delta^{(m-1)}(x - x_0) \\
&= + \cdots + g^{(m)}(x_0)\delta(x - x_0) + \theta(x - x_0)g^{(m+1)}(x).
\end{aligned}$$

This infers that for $x \geq x_0$

$$\begin{aligned}
f(x) &= \frac{d^{m+1}}{dx^{m+1}}\theta(x - x_0)g(x) * \frac{x_+^{-\alpha+m}}{\Gamma(-\alpha + m + 1)} \\
&= g(x_0)\frac{(x - x_0)_+^{-\alpha}}{\Gamma(-\alpha + 1)} + \cdots + g^{(m)}(x_0)\frac{(x - x_0)_+^{-\alpha+m}}{\Gamma(-\alpha + m + 1)} \\
&\quad + \frac{1}{\Gamma(-\alpha + m + 1)} \int_{x_0}^x g^{(m+1)}(\zeta)(x - \zeta)^{-\alpha+m}d\zeta.
\end{aligned}$$

The rest of the proof follows easily. This completes the proof of Theorem 1. $\square$

Example 2. *Let* $m = 0, 1, \ldots$. *Then the generalized Abel's integral equation*

$$\theta(x-1)e^x = \int_0^x f(\tau)(x-\tau)^{m-0.5}d\tau \tag{5}$$

has the solution in $\mathcal{D}'(R^+)$

$$
\begin{aligned}
f(x) = \ & \frac{1}{\binom{m-0.5}{m}m!\sqrt{\pi}}\Big\{\delta^{(m-0.5)}(x-1) \\
& + \delta^{(m-3/2)}(x-1) + \cdots + \delta^{(-0.5)}(x-1) + \frac{1}{\sqrt{\pi}}\int_1^x \theta(\tau-1)e^\tau(x-\tau)^{-0.5}d\tau\Big\}.
\end{aligned}
$$

Proof. Clearly, we can write Equation (5) to

$$\theta(x-1)e^x = \frac{\Gamma(m+0.5)}{\Gamma(m+0.5)}\int_0^x f(\tau)(x-\tau)^{m-0.5}d\tau = \Gamma(m+0.5)\Phi_{m+0.5} * f,$$

which deduces that by Theorem 1

$$f(x) = \frac{1}{\Gamma(m+0.5)}\Phi_{-m-0.5} * \theta(x-1)e^x = \frac{1}{\Gamma(m+0.5)}\Phi_{0.5} * \Phi_{-m-1} * \theta(x-1)e^x$$

as

$$\Phi_0 * f = \delta * f = f.$$

Obviously,

$$\left(\frac{d}{dx}\theta(x-1)e^x, \phi(x)\right) = -\int_1^\infty e^x \phi'(x)dx = \phi(1) + \int_{-\infty}^\infty \theta(x-1)e^x\phi(x)dx$$

which claims that

$$\frac{d}{dx}\theta(x-1)e^x = \delta(x-1) + \theta(x-1)e^x.$$

Similarly, we have

$$\frac{d^2}{dx^2}\theta(x-1)e^x = \delta'(x-1) + \delta(x-1) + \theta(x-1)e^x.$$

By mathematical induction, we come to

$$\frac{d^{m+1}}{dx^{m+1}}\theta(x-1)e^x = \delta^{(m)}(x-1) + \delta^{(m-1)}(x-1) + \cdots + \delta(x-1) + \theta(x-1)e^x.$$

Therefore,

$$
\begin{aligned}
& \Phi_{0.5} * \Phi_{-m-1} * \theta(x-1)e^x \\
& = \Phi_{0.5} * \frac{d^{m+1}}{dx^{m+1}}\theta(x-1)e^x \\
& = \Phi_{0.5} * \Big\{\delta^{(m)}(x-1) + \delta^{(m-1)}(x-1) + \cdots + \delta(x-1) + \theta(x-1)e^x\Big\} \\
& = \delta^{(m-0.5)}(x-1) + \delta^{(m-3/2)}(x-1) + \cdots + \delta^{(-0.5)}(x-1) + \frac{1}{\sqrt{\pi}}\int_1^x \theta(\tau-1)e^\tau(x-\tau)^{-0.5}d\tau.
\end{aligned}
$$

This completes the proof by noting that

$$\Gamma(m+0.5) = \binom{m-0.5}{m}m!\sqrt{\pi}.$$

□

Example 3. *Let $m = 1, 2, \ldots$ and*

$$\sin x_+ = \begin{cases} \sin x & \text{if } x \geq 0, \\ 0, & \text{otherwise.} \end{cases}$$

Then the generalized Abel's integral equation

$$\sin x_+ = \int_0^x f(\tau)(x - \tau)^{-m-0.5} d\tau \tag{6}$$

has the solution in $\mathcal{D}'(R^+)$

$$f(x) = \frac{\binom{-0.5}{m} m! \, x_+^{m+0.5}}{\sqrt{\pi}} E_{2,m+3/2}(-x^2)$$

where $E_{\alpha,\beta}(x)$ is the Mittag-Leffler function, defined by the series expansion

$$E_{\alpha,\beta}(x) = \sum_{k=0}^{\infty} \frac{x^k}{\Gamma(\alpha k + \beta)}$$

for α, $\beta > 0$.

Proof. Evidently, we can convert Equation (6) to

$$\sin x_+ = \frac{\Gamma(-m+0.5)}{\Gamma(-m+0.5)} \int_0^x f(\tau)(x-\tau)^{-m-0.5} d\tau = \Gamma(-m+0.5)\Phi_{-m+0.5} * f.$$

Hence from Theorem 1, we get

$$\begin{aligned}
f(x) &= \frac{1}{\Gamma(-m+0.5)} \Phi_{m-0.5} * \sin x_+ = \frac{1}{\Gamma(-m+0.5)\Gamma(m-0.5)} \int_0^x \sin \tau \, (x-\tau)^{m-3/2} d\tau \\
&= \frac{1}{\Gamma(-m+0.5)\Gamma(m-0.5)} \sum_{k=0}^{\infty} \frac{(-1)^k}{(2k+1)!} \int_0^x \tau^{2k+1} (x-\tau)^{m-3/2} d\tau.
\end{aligned}$$

Setting $\tau = xt$, we arrive at

$$\int_0^x \tau^{2k+1}(x-\tau)^{m-3/2} d\tau = \frac{\Gamma(2k+2)\Gamma(m-0.5)}{\Gamma(2k+m+3/2)} x_+^{2k+m+0.5}.$$

This implies

$$\begin{aligned}
f(x) &= \frac{1}{\Gamma(-m+0.5)} \sum_{k=0}^{\infty} \frac{(-1)^k}{\Gamma(2k+m+3/2)} x_+^{2k+m+0.5} \\
&= \frac{x_+^{m+0.5}}{\Gamma(-m+0.5)} \sum_{k=0}^{\infty} \frac{(-x^2)^k}{\Gamma(2k+m+3/2)} \\
&= \frac{x_+^{m+0.5}}{\Gamma(-m+0.5)} E_{2,m+3/2}(-x^2).
\end{aligned}$$

This completes the proof by using

$$\Gamma(-m+0.5) = \frac{\sqrt{\pi}}{\binom{-0.5}{m} m!}.$$

Furthermore, the generalized Abel's integral equation

$$\delta^{(k)}(x) + x_+^{-\sqrt{3}} = \int_0^x f(\tau)(x-\tau)^{-m-0.5}d\tau$$

has the solution in $\mathcal{D}'(R^+)$

$$f(x) = \frac{\binom{-0.5}{m}m!}{\sqrt{\pi}}\Phi_{m-k-0.5}(x) + \frac{\binom{-0.5}{m}\Gamma(-\sqrt{3}+1)\,m!}{\sqrt{\pi}}\Phi_{m-\sqrt{3}+0.5}(x)$$

for $k = 0, 1, \ldots$ and $m = 1, 2, \ldots$.

Similarly, the generalized Abel's integral equation

$$\delta^{(k)}(x) + x_+^{-\sqrt{3}} = \int_0^x f(\tau)(x-\tau)^{m-0.5}d\tau \tag{7}$$

has the solution in $\mathcal{D}'(R^+)$

$$f(x) = \frac{1}{\binom{m-0.5}{m}m!\sqrt{\pi}}\Phi_{-m-k-0.5}(x) + \frac{\Gamma(-\sqrt{3}+1)}{\binom{m-0.5}{m}m!\sqrt{\pi}}\Phi_{-m-\sqrt{3}+0.5}(x).$$

for $k, m = 0, 1, \ldots$. □

To end off this section, we would add that many applied problems from physical, engineering and chemical processes lead to integral equations, which at first glance have nothing in common with Abel's integral equations, and due to this perception, additional efforts are undertaken for the development of analytical or numerical procedure for solving these equations. However, their transformations to the form of Abel's integral equations will speed up the solution process [24], or, more significantly, lead to distributional solutions in cases where classical ones do not exist [22].

As an example, the following integral equation with a moving integration limit

$$\int_0^y \frac{1}{(y^2-x^2)^\beta}f(x)dx = g(y)$$

can be converted into Abel's integral equation to solve in the distributional sense. The interested readers are referred to [22] for the detailed methods.

3. Babenko's Approach in Distribution

In this section, we shall extend the method used by Yu. I. Babenko in his book [28], for solving various types of fractional differential and integral equations in the classical sense, to generalized functions. The method itself is close to the Laplace transform method in the ordinary sense, but it can be used in more cases, such as solving fractional differential equations with variable coefficients. Clearly, it is always necessary to show convergence of the series obtained as solutions by other analytical tools, although it is a hard job in general [24]. We point out that Babenko's method can also be used to solve certain partial differential equations for heat and mass transfer theory, and suggest the interested readers are referred to [24] for the detailed arguments.

We must add that Oliver Heaviside (1850–1925) introduced an ingenious way of solving ordinary or partial differential equations arising from electromagnetic problems through algebration [29–31]. He considered, for example, the factor $d^2/dx^2 f$ as product of the differential operator d^2/dx^2 with f itself. Therefore, the differential equation for a given $g(x)$

$$y(x) + y''(x) = g(x)$$

can be converted to

$$(1 + d^2/dx^2)y(x) = g(x),$$

which implies that the solution

$$y(x) = (1 + d^2/dx^2)^{-1}g(x) = \sum_{n=0}^{\infty}(-1)^n D^{2n}g(x), \quad D = d/dx$$

if it converges. This method is identical to Babenko's Approach.

Let $g(x) \in \mathcal{D}'(R^+)$ be given. We now study Abel's integral equation of the second kind

$$y(x) + \frac{\lambda}{\Gamma(\alpha)}\int_0^x (x - \tau)^{\alpha-1} y(\tau)\, d\tau = g(x) \tag{8}$$

with demonstrations of examples in the distributional space $\mathcal{D}'(R^+)$, where $\alpha \in R$ and λ is a constant. Further, we will investigate and solve several integral equations with variable coefficients by products of distributions and fractional operations of generalized functions. The results derived here cannot be achieved by the Laplace transform in general, or numerical analysis methods since distributions are undefined at points in R.

Clearly, Equation (8) is equivalent to the convolutional equation

$$(\delta + \lambda\Phi_\alpha) * y(x) = g(x)$$

in $\mathcal{D}'(R^+)$, although it is undefined in the classical sense for $\alpha \leq 0$. We should point out it becomes the differential equation

$$y + \lambda y * \delta^{(m)} = y + \lambda y^{(m)} = g$$

if $\alpha = -m$.

Example 4. *Let λ be a nonzero constant and $\alpha > 0$. Then the fractional differential equation*

$$\lambda y(x) + y^{(\alpha)}(x) = \delta(x) \tag{9}$$

has the solution in the space $\mathcal{D}'(R^+)$

$$y(x) = x_+^{\alpha-1} E_{\alpha,\alpha}(-\lambda x_+^\alpha).$$

Proof. Equation (9) can be written into

$$\lambda y + \Phi_{-\alpha} * y = \delta$$

which is equivalent to

$$\lambda \Phi_\alpha * y + y = \Phi_\alpha * \delta.$$

By Babenko's method we get

$$
\begin{aligned}
y(x) &= \sum_{n=0}^{\infty}(-1)^n \lambda^n \Phi_\alpha^{n+1} * \delta = \sum_{n=0}^{\infty}(-1)^n \lambda^n \Phi_\alpha^{n+1} \\
&= \sum_{n=0}^{\infty}(-1)^n \lambda^n \Phi_{\alpha(n+1)} = \sum_{n=0}^{\infty}(-1)^n \lambda^n \frac{x_+^{\alpha n + \alpha - 1}}{\Gamma(\alpha n + \alpha)} \\
&= x_+^{\alpha-1} \sum_{n=0}^{\infty}\frac{(-\lambda x_+^\alpha)^n}{\Gamma(\alpha n + \alpha)} = x_+^{\alpha-1} E_{\alpha,\alpha}(-\lambda x_+^\alpha)
\end{aligned}
$$

which is convergent, and hence well defined, by noting that $x_+^{\alpha-1}$ is a locally integrable function on R.

In particular, the differential equation

$$\lambda y(x) + y'(x) = \delta(x)$$

has the solution in the space $\mathcal{D}'(R^+)$

$$y(x) = \theta(x)e^{-\lambda x}.$$

Similarly, the fractional differential equation

$$\lambda y(x) + y^{(\alpha)}(x) = x_+$$

has the solution

$$y(x) = x_+^{\alpha+1}E_{\alpha,\alpha+2}(-\lambda x_+^\alpha)$$

where $\alpha > 0$.

Indeed,

$$
\begin{aligned}
y(x) &= \sum_{n=0}^{\infty}(-1)^n\lambda^n\Phi_\alpha^{n+1} * x_+ = \sum_{n=0}^{\infty}(-1)^n\lambda^n\Phi_\alpha^{n+1} * \frac{x_+}{\Gamma(2)} \\
&= \sum_{n=0}^{\infty}(-1)^n\lambda^n\Phi_{(n+1)\alpha} * \Phi_2 = \sum_{n=0}^{\infty}(-1)^n\lambda^n\Phi_{(n+1)\alpha+2}(x) \\
&= x_+^{\alpha+1}\sum_{n=0}^{\infty}(-1)^n\lambda^n\frac{x_+^{\alpha n}}{\Gamma(n\alpha+\alpha+2)} = x_+^{\alpha+1}E_{\alpha,\alpha+2}(-\lambda x_+^\alpha).
\end{aligned}
$$

Using the same argument, the fractional differential equation

$$\lambda y(x) + y^{(\alpha)}(x) = \delta'(x)$$

has the solution

$$y(x) = x_+^{\alpha-2}E_{\alpha,\alpha-1}(-\lambda x_+^\alpha)$$

where $\alpha > 1$ and $x_+^{\alpha-2}$ is a regular distribution.

More generally, the fractional differential equation

$$y^{(\alpha)}(x) + \lambda y^{(\beta)}(x) = \delta(x), \quad \alpha > \beta \geq 0$$

has the solution

$$y(x) = x_+^{\alpha-1}E_{\alpha-\beta,\alpha}(-\lambda x_+^{\alpha-\beta}).$$

□

Remark 1. *We begin by using Example 4 as a simple demonstration of Babenko's Approach. Clearly, all the equations presented above can be easily solved by the Laplace transform. For example, applying the Laplace transform to the equation*

$$y^{(\alpha)}(x) + \lambda y^{(\beta)}(x) = \delta(x), \quad \alpha > \beta \geq 0$$

gives

$$y^*(s) = \frac{1}{s^\alpha + \lambda s^\beta}.$$

The inverse transform implies

$$y(x) = x_+^{\alpha-1}E_{\alpha-\beta,\alpha}(-\lambda x_+^{\alpha-\beta}).$$

However, we must mention that Babenko's approach in the distributional sense is much more general than that of the Laplace transform. As an example, the Laplace transform does not work for the equation

$$y^{(\alpha)}(x) + \lambda y^{(\beta)}(x) = x_+^{-1.5}, \quad \alpha > \beta \geq 0$$

as the distribution $x_+^{-1.5}$ is not locally integrable. As indicated below, Babenko's approach provides an efficient method of dealing with this kind of equation.

Example 5. *Let $\alpha > 0$. Then the fractional differential and integral equation (mixed type)*

$$y^{(0.75)}(x) + \frac{\lambda}{\Gamma(\alpha)} \int_0^x (x-\tau)^{\alpha-1} y(\tau) d\tau = x_+^{-1.5} \tag{10}$$

has the solution in the space $\mathcal{D}'(R^+)$

$$y(x) = -2\sqrt{\pi} x_+^{-0.75} E_{\alpha+0.75,\,0.25}(-\lambda x_+^{\alpha+0.75}).$$

Proof. Equation (10) can be changed to

$$y^{(0.75)}(x) + \lambda y^{(-\alpha)}(x) = \Gamma(-0.5)\Phi_{-0.5}(x)$$

which is equivalent to

$$\Phi_{-0.75} * y + \lambda \Phi_\alpha * y = -2\sqrt{\pi}\Phi_{-0.5}.$$

Applying $\Phi_{0.75}$ to both sides of the above equation, we only need to study the following integral equation

$$y + \lambda \Phi_{\alpha+0.75} * y = -2\sqrt{\pi}\Phi_{0.25}.$$

Therefore,

$$
\begin{aligned}
y(x) &= -2\sqrt{\pi} \sum_{n=0}^{\infty} (-1)^n \lambda^n \Phi_{\alpha n+0.75n} * \Phi_{0.25} = -2\sqrt{\pi} \sum_{n=0}^{\infty} (-1)^n \lambda^n \Phi_{\alpha n+0.75n+0.25} \\
&= -2\sqrt{\pi} \sum_{n=0}^{\infty} (-1)^n \lambda^n \frac{x_+^{\alpha n+0.75n-0.75}}{\Gamma(\alpha n+0.75n+0.25)} \\
&= -2\sqrt{\pi} x_+^{-0.75} \sum_{n=0}^{\infty} \frac{(-\lambda x_+^{\alpha+0.75})^n}{\Gamma(\alpha n+0.75n+0.25)} \\
&= -2\sqrt{\pi} x_+^{-0.75} E_{\alpha+0.75,\,0.25}(-\lambda x_+^{\alpha+0.75})
\end{aligned}
$$

which is convergent. □

Example 6. *The generalized Abel's integral equation*

$$y(x) + \int_0^x \frac{y(\tau)}{\sqrt{x-\tau}} d\tau = x_+^{-\sqrt{2}} + \delta(x) \tag{11}$$

has the solution in the space $\mathcal{D}'(R^+)$

$$
\begin{aligned}
y(x) &= x_+^{-\sqrt{2}} + \delta(x) - x_+^{-0.5} - \frac{\Gamma(1-\sqrt{2})\sqrt{\pi}\, x_+^{0.5-\sqrt{2}}}{\Gamma(3/2-\sqrt{2})} + \frac{\pi}{1-\sqrt{2}} x_+^{1-\sqrt{2}} \\
&\quad - \Gamma(1-\sqrt{2})\pi^{3/2} x_+^{3/2-\sqrt{2}} E_{0.5,\,5/2-\sqrt{2}}(-\sqrt{\pi} x_+^{0.5}) + \pi E_{0.5,\,1}(-\sqrt{\pi} t_+^{0.5}).
\end{aligned}
$$

Proof. Equation (11) can be written into

$$y(x) + \frac{\sqrt{\pi}}{\Gamma(0.5)} \int_0^x \frac{y(\tau)}{\sqrt{x-\tau}} \, d\tau = \Gamma(1-\sqrt{\pi}) \frac{x_+^{-\sqrt{2}}}{\Gamma(1-\sqrt{\pi})} + \frac{x_+^{-1}}{\Gamma(0)} = \Gamma(1-\sqrt{\pi})\Phi_{1-\sqrt{\pi}} + \Phi_0.$$

By Babenko's method we get

$$\begin{aligned}
y(x) &= \sum_{n=0}^{\infty} (-1)^n \pi^{n/2} \Phi_{n/2} * \left(\Gamma(1-\sqrt{\pi})\Phi_{1-\sqrt{\pi}} + \Phi_0 \right) \\
&= \sum_{n=0}^{\infty} (-1)^n \pi^{n/2} \Phi_{n/2} * \Gamma(1-\sqrt{\pi})\Phi_{1-\sqrt{\pi}} \\
&\quad + \sum_{n=0}^{\infty} (-1)^n \pi^{n/2} \Phi_{n/2} \triangleq I_1 + I_2.
\end{aligned}$$

Clearly,

$$\begin{aligned}
I_2 &= \sum_{n=0}^{\infty} (-1)^n \pi^{n/2} \Phi_{n/2} = \sum_{n=0}^{\infty} (-1)^n \pi^{n/2} \frac{x_+^{n/2-1}}{\Gamma(n/2)} \\
&= \delta(x) - \sqrt{\pi} \frac{x_+^{-0.5}}{\Gamma(0.5)} + \sum_{n=2}^{\infty} (-1)^n \pi^{n/2} \frac{x_+^{n/2-1}}{\Gamma(n/2)} \\
&= \delta(x) - x_+^{-0.5} + \pi E_{0.5,1}(-\sqrt{\pi} x_+^{0.5})
\end{aligned}$$

where

$$E_{0.5,1}(x) = e^{x^2} \operatorname{erfc}(-x)$$

and $\operatorname{erfc}(x)$ is the error function complement defined by

$$\operatorname{erfc}(x) = \frac{2}{\sqrt{\pi}} \int_x^{\infty} e^{-t^2} \, dt.$$

As for I_1, we come to

$$\begin{aligned}
I_1 &= \Gamma(1-\sqrt{2}) \sum_{n=0}^{\infty} (-1)^n \pi^{n/2} \Phi_{n/2+1-\sqrt{2}} \\
&= x_+^{-\sqrt{2}} - \frac{\Gamma(1-\sqrt{2})\sqrt{\pi}\, x_+^{0.5-\sqrt{2}}}{\Gamma(3/2-\sqrt{2})} \\
&\quad + \Gamma(1-\sqrt{2})\pi \frac{x_+^{1-\sqrt{2}}}{\Gamma(2-\sqrt{2})} + \Gamma(1-\sqrt{2}) \sum_{n=3}^{\infty} (-1)^n \pi^{n/2} \Phi_{n/2+1-\sqrt{2}} \\
&= x_+^{-\sqrt{2}} - \frac{\Gamma(1-\sqrt{2})\sqrt{\pi}\, x_+^{0.5-\sqrt{2}}}{\Gamma(3/2-\sqrt{2})} + \frac{\pi}{1-\sqrt{2}} x_+^{1-\sqrt{2}} \\
&\quad - \Gamma(1-\sqrt{2})\pi^{3/2} x_+^{3/2-\sqrt{2}} E_{0.5,5/2-\sqrt{2}}(-\sqrt{\pi} x_+^{0.5}).
\end{aligned}$$

The result follows from the sum of I_1 and I_2. $\quad\square$

Remark 2. *Equation* (11) *cannot be discussed in the classical sense, including the Laplace transform, since the fractional integral or derivative of* $x_+^{-\sqrt{2}}$, *does not exist in the normal sense. Clearly, the distribution* $x_+^{-\sqrt{2}} + \delta(x)$ *in the solution is a singular generalized function in* $\mathcal{D}'(R^+)$, *while the rest*

$$x_+^{-0.5} - \frac{\Gamma(1-\sqrt{2})\sqrt{\pi}\,x_+^{0.5-\sqrt{2}}}{\Gamma(3/2-\sqrt{2})} + \frac{\pi}{1-\sqrt{2}}x_+^{1-\sqrt{2}}$$
$$-\Gamma(1-\sqrt{2})\pi^{3/2}x_+^{3/2-\sqrt{2}}E_{0.5,5/2-\sqrt{2}}(-\sqrt{\pi}x_+^{0.5}) + \pi E_{0.5,1}(-\sqrt{\pi}x_+^{0.5}).$$

is regular (locally integrable).

Assume f is a distribution in $\mathcal{D}'(R)$ and g is a function in $C^\infty(R)$. Then the product fg is well defined by

$$(fg, \phi) = (f, g\phi)$$

for all functions $\phi \in \mathcal{D}(R)$ as $g\phi \in \mathcal{D}(R)$. Therefore, the product, for $k = 0, 1, \ldots,$

$$\frac{\lambda x^k}{\Gamma(\alpha)} \int_0^x (x-\tau)^{\alpha-1} y(\tau)\, d\tau$$

makes sense since x^k is in $C^\infty(R)$ and

$$\frac{\lambda}{\Gamma(\alpha)} \int_0^x (x-\tau)^{\alpha-1} y(\tau)\, d\tau = \lambda \Phi_\alpha * y$$

is a distribution in $\mathcal{D}'(R^+)$ (subspace of $\mathcal{D}'(R)$) for arbitrary $\alpha \in R$ if $y \in \mathcal{D}'(R^+)$, according to Section 2.

Example 7. *Let $k \geq 1$ be an integer. Then the integral equation*

$$x^k \int_0^x \frac{y(\tau)}{\sqrt{x-\tau}}\, d\tau = \delta^{(m)}(x)$$

has a solution in the space $\mathcal{D}'(R^+)$

$$y(x) = \frac{1}{\sqrt{\pi}}\left\{(-1)^k \frac{\delta^{(m+k+0.5)}(x)}{k!\binom{m+k}{k}} + C_0\delta^{(0.5)}(x) + \cdots + C_{k-1}\delta^{(k-0.5)}(x)\right\}$$

where m is a non-negative integer and $C_0, \cdots, C_{k-1}$ are arbitrary constants.

Proof. First, we show that

$$x^k \delta^{(s)}(x) = \begin{cases} (-1)^k k!\binom{s}{k}\delta^{(s-k)}(x) & \text{if } k \leq s, \\ 0 & \text{otherwise} \end{cases}$$

for $k, s = 0, 1, 2, \ldots.$

Indeed,

$$\begin{aligned}(x^k\delta^{(s)}(x), \phi(x)) &= (-1)^s(x^k\phi(x))^{(s)}\big|_{x=0} \\ &= (-1)^s k!\binom{s}{k}\phi^{(s-k)}(0) \\ &= (-1)^k k!\binom{s}{k}(\delta^{(s-k)}(x), \phi(x))\end{aligned}$$

for $k \leq s.$

On the other hand, we have

$$x^k \delta^{(s)}(x) = 0$$

if $k > s$.

It follows that

$$x^k \, \delta^{(m+k)}(x) = (-1)^k k! \binom{m+k}{k} \delta^{(m)}(x).$$

Therefore,

$$\int_0^x \frac{y(\tau)}{\sqrt{x-\tau}} \, d\tau = (-1)^k \frac{\delta^{(m+k)}(x)}{k! \binom{m+k}{k}} + C_0 \delta(x) + \cdots + C_{k-1} \delta^{(k-1)}(x)$$

which implies that

$$
\begin{aligned}
y(x) &= \frac{1}{\sqrt{\pi}} \Phi_{-0.5} * \left\{ (-1)^k \frac{\delta^{(m+k)}(x)}{k! \binom{m+k}{k}} + C_0 \delta(x) + \cdots + C_{k-1} \delta^{(k-1)}(x) \right\} \\
&= \frac{1}{\sqrt{\pi}} \left\{ (-1)^k \frac{\delta^{(m+k+0.5)}(x)}{k! \binom{m+k}{k}} + C_0 \delta^{(0.5)}(x) + \cdots + C_{k-1} \delta^{(k-0.5)}(x) \right\}
\end{aligned}
$$

where $C_0, \cdots, C_{k-1}$ are arbitrary constants.

Furthermore, we let $k \geq 1$ be an integer. Then the integral equation

$$x^k \int_0^x \frac{y(\tau)}{(x-\tau)^{3/2}} \, d\tau = \delta^{(m)}(x)$$

has a solution in the space $\mathcal{D}'(R^+)$

$$y(x) = -2\sqrt{\pi} \left\{ (-1)^k \frac{\delta^{(m+k-0.5)}(x)}{k! \binom{m+k}{k}} + C_0 \delta^{(-0.5)}(x) + \cdots + C_{k-1} \delta^{(k-3/2)}(x) \right\}$$

where m is a non-negative integer and $C_0, \cdots, C_{k-1}$ are arbitrary constants.

In fact, we derive that

$$
\begin{aligned}
y(x) &= \Gamma(-0.5) \Phi_{0.5} * \left\{ (-1)^k \frac{\delta^{(m+k)}(x)}{k! \binom{m+k}{k}} + C_0 \delta(x) + \cdots + C_{k-1} \delta^{(k-1)}(x) \right\} \\
&= -2\sqrt{\pi} \left\{ (-1)^k \frac{\delta^{(m+k-0.5)}(x)}{k! \binom{m+k}{k}} + C_0 \delta^{(-0.5)}(x) + \cdots + C_{k-1} \delta^{(k-3/2)}(x) \right\}.
\end{aligned}
$$

It generally follows that the integral equation

$$x^k \int_0^x y(\tau)(x-\tau)^{\alpha-1} \, d\tau = \delta^{(m)}(x)$$

has a solution

$$y(x) = \frac{1}{\Gamma(\alpha)} \left\{ (-1)^k \frac{\delta^{(m+k+\alpha)}(x)}{k! \binom{m+k}{k}} + C_0 \delta^{(\alpha)}(x) + \cdots + C_{k-1} \delta^{(k+\alpha-1)}(x) \right\}$$

where $\alpha \neq 0, -1, \ldots$. $\square$

.

Example 8. *Let $k \geq 1$ be an integer. Then the generalized integral equation*

$$x_+^k \int_0^x \frac{y(\tau)}{\sqrt{x-\tau}} d\tau = \delta^{(m)}(x) \tag{12}$$

has a solution in the space $\mathcal{D}'(R^+)$

$$y(x) = \frac{1}{\sqrt{\pi}} \left\{ \frac{2(-1)^k m!}{(m+k)!} \delta^{(k+m+0.5)}(x) + C_0 \delta^{(0.5)}(x) + \cdots + C_{k-1} \delta^{(k-0.5)}(x) \right\}.$$

where $m = 0, 1, 2, \ldots$ and $C_0, \cdots, C_{k-1}$ are arbitrary constants.

Proof. To solve this integral equation, we require the following more complicated product of distributions, as x_+^k is not an infinitely differentiable function (but it is a locally integrable function). It follows from Theorem 2.3 in [32] that

$$x_+^r \delta^{(r+m)}(x) = \frac{(-1)^r (r+m)!}{2m!} \delta^{(m)}(x)$$

for $m, r = 0, 1, 2, \ldots$. We note that this product can be derived directly from the complex analysis approach based on the Laurent series of x_+^λ, x_-^λ as well as $e^{\pm i\lambda\pi}$, given below

$$x_+^\lambda = \frac{(-1)^{n-1}}{(n-1)!(\lambda+n)} \delta^{(n-1)}(x) + F_{-n}(x_+, \lambda),$$

$$x_-^\lambda = \frac{1}{(n-1)!(\lambda+n)} \delta^{(n-1)}(x) + F_{-n}(x_-, \lambda),$$

$$e^{\pm i\lambda\pi} = (-1)^n [1 \pm (\lambda+n)\pi + \cdots].$$

Hence,

$$\int_0^x \frac{y(\tau)}{\sqrt{x-\tau}} d\tau = \frac{2(-1)^k m!}{(m+k)!} \delta^{(k+m)}(x) + C_0 \delta(x) + \cdots + C_{k-1}\delta^{(k-1)}(x)$$

by noting that

$$x_+^k \delta^{(i)}(x) = 0$$

where $i = 0, 1, 2, \cdots, k-1$ and $C_0, \cdots, C_{k-1}$ are arbitrary constants. This implies that Equation (12) has a solution

$$\begin{aligned} y(x) &= \frac{1}{\sqrt{\pi}} \Phi_{-0.5} * \left\{ \frac{2(-1)^k m!}{(m+k)!} \delta^{(k+m)}(x) + C_0 \delta(x) + \cdots + C_{k-1}\delta^{(k-1)}(x) \right\} \\ &= \frac{1}{\sqrt{\pi}} \left\{ \frac{2(-1)^k m!}{(m+k)!} \delta^{(k+m+0.5)}(x) + C_0 \delta^{(0.5)}(x) + \cdots + C_{k-1}\delta^{(k-0.5)}(x) \right\}. \end{aligned}$$

Similarly, we let $k \geq 1$ be an integer. Then, the integral equation

$$x_+^k \int_0^x \frac{y(\tau)}{(x-\tau)^{3/2}} d\tau = \delta^{(m)}(x)$$

has a solution in the space $\mathcal{D}'(R^+)$

$$y(x) = -\frac{1}{2\sqrt{\pi}} \left\{ \frac{2(-1)^k m!}{(m+k)!} \delta^{(k+m-0.5)}(x) + C_0 \delta^{(-0.5)}(x) + \cdots + C_{k-1}\delta^{(k-3/2)}(x) \right\}$$

where m is a non-negative integer, and $C_0, \cdots, C_{k-1}$ are arbitrary constants.

Clearly, the integral equation

$$x_+^k \int_0^x y(\tau)(x-\tau)^{\alpha-1} \, d\tau = \delta^{(m)}(x)$$

has a solution

$$y(x) = \frac{1}{\Gamma(\alpha)} \left\{ \frac{2(-1)^k m!}{(m+k)!} \delta^{(k+m+\alpha)}(x) + C_0 \delta^{(\alpha)}(x) + \cdots + C_{k-1} \delta^{(k+\alpha-1)}(x) \right\}$$

where $\alpha \neq 0, -1, \ldots$. $\square$

Gorenflo and Mainardi [7] presented the applications of Abel's integral equations of the first and second kind to solve the following partial equation of heat flow:

$$u_t - u_{xx} = 0, \quad u = u(x,t)$$

in the semi-infinite intervals $0 < x < \infty$ and $0 < t < \infty$ of space and time, respectively. Our results on the distributional Abel's integral equations have potential applications to dealing with differential equations in distribution and hence finding distributional (weak) solutions.

Remark 3. *Generally speaking, there is the lack of definitions for nonlinear operations, such as product and composition in distribution theory, although it is of great demand in the areas of differential equations and quantum field theory. As an example, it seems hard to define the distribution $\delta^2(x)$, as*

$$(\delta^2(x), \phi(x)) = (\delta(x), \delta(x)\phi(x)) = \delta(0)\phi(0)$$

is undefined. Fisher, with his coauthors [33–41], has actively used the δ-sequence and neutrix limit due to van der Corput since 1969, to deduce numerous products, powers, convolutions, and compositions of distributions by several workable definitions. From the above examples, it is clear to see the relations between products of generalized functions and integral equations with variable coefficients in distribution.

4. Conclusions

Applying Babenko's method and fractional calculus, we have studied and solved several fractional differential and integral equations, including ones with variable coefficients and a mixed type equation, in the distributional space $\mathcal{D}'(R^+)$ by using well defined products of generalized functions and the Mittag-Leffler function. In particular, we discussed Abel's integral equations of the first and second kind for arbitrary $\alpha \in R$ by fractional operations of distributions, and derived several new and interesting results which cannot be realized by numerical analysis, as they involve distributions which are undefined on R, or by the Laplace transform.

Author Contributions: The order of the author list reflects contributions to the paper.

Funding: This research was funded by NSERC (Canada) under grant number 2017-00001 and NSFC (China) under grant number 11671251.

Acknowledgments: The authors are grateful to the reviewers for the careful reading of the paper with several productive suggestions and corrections, which certainly improved its quality.

Conflicts of Interest: The authors declare no conflict of interest.

References

1. Zimbardo, G.; Perri, S.; Effenberger, F.; Fichtner, H. Fractional Parker equation for the transport of cosmic rays: Steady-state solutions. *Astron. Astrophys.* **2017**, *607*, A7. [CrossRef]

2. Li, C.; Li, C.P. On defining the distributions $(\delta)^k$ and $(\delta')^k$ by fractional derivatives. *Appl. Math. Comput.* **2014**, *246*, 502–513.

3. Mann, W.R.; Wolf, F. Heat transfer between solids and gases under nonlinear boundary conditions. *Quart. Appl. Math.* **1951**, *9*, 163–184. [CrossRef]

4. Goncerzewicz, J.; Marcinkowska, H.; Okrasinski, W.; Tabisz, K. On percolation of water from a cylindrical reservoir into the surrounding soil. *Appl. Math.* **1978**, *16*, 249–261. [CrossRef]

5. Keller, J.J. Propagation of simple nonlinear waves in gas filled tubes with friction. *Z. Angew. Math. Phys.* **1981**, *32*, 170–181. [CrossRef]

6. Kilbas, A.A.; Srivastava, H.M.; Trujillo, J.J. *Theory and Applications of Fractional Differential Equations*; Elsevier: New York, NY, USA, 2006.

7. Gorenflo, R.; Mainardi, F. Fractional Calculus: Integral and Differential Equations of Fractional Order. In *Fractals and Fractional Calculus in Continuum Mechanics*; Carpinteri, A., Mainardi, F., Eds.; Springer: New York, NY, USA, 1997; pp. 223–276.

8. Wang, J.R.; Zhu, C.; Fečkan, M. Analysis of Abel-type nonlinear integral equations with weakly singular kernels. *Bound. Value Probl.* **2014**, *2014*, 20. [CrossRef]

9. Atkinson, K.E. An existence theorem for Abel integral equations. *SIAM J. Math. Anal.* **1974**, *5*, 729–736. [CrossRef]

10. Bushell, P.J.; Okrasinski, W. Nonlinear Volterra integral equations with convolution kernel. *J. Lond. Math. Soc.* **1990**, *41*, 503–510. [CrossRef]

11. Gorenflo, R.; Vessella, S. *Abel Integral Equations: Analysis and Applications*; Springer-Verlag: Berlin, Germany, 1991; Volume 1461.

12. Okrasinski, W. Nontrivial solutions to nonlinear Volterra integral equations. *SIAM J. Math. Anal.* **1991**, *22*, 1007–1015. [CrossRef]

13. Gripenberg, G. On the uniqueness of solutions of Volterra equations. *J. Integral Equ. Appl.* **1990**, *2*, 421–430. [CrossRef]

14. Mydlarczyk, W. The existence of nontrivial solutions of Volterra equations. *Math. Scand.* **1991**, *68*, 83–88. [CrossRef]

15. Kilbas, A.A.; Saigo, M. On solution of nonlinear Abel-Volterra integral equation. *J. Math. Anal. Appl.* **1999**, *229*, 41–60. [CrossRef]

16. Karapetyants, N.K.; Kilbas, A.A.; Saigo, M. Upper and lower bounds for solutions of nonlinear Volterra convolution integral equations with power nonlinearity. *J. Integral Equ. Appl.* **2000**, *12*, 421–448. [CrossRef]

17. Lima, P.; Diogo, T. Numerical solution of nonuniquely solvable Volterra integral equation using extrapolation methods. *J. Comput. App. Math.* **2002**, *140*, 537–557. [CrossRef]

18. Rahimy, M. Applications of fractional differential equations. *Appl. Math. Sci.* **2010**, *4*, 2453–2461.

19. Li, C.P.; Zeng, F. *Numerical Methods for Fractional Calculus*; Chapman and Hall/CRC: Boca Raton, FL, USA, 2015.

20. Mainardi, F. Fractional relaxation-oscillation and fractional diffusion-wave phenomena. *Chaos Solitons Fractals* **1996**, *7*, 1461–1477. [CrossRef]

21. Kesavan, S. *Functional Analysis*; Hindustan Book Agency: Gurgaon, India, 2014.

22. Li, C.; Li, C.P.; Kacsmar, B.; Lacroix, R.; Tilbury, K. The Abel integral equations in distributions. *Adv. Anal.* **2017**, *2*, 88–104. [CrossRef]

23. Li, C.; Clarkson, K. Babenko's approach to Abel's integral equations. *Mathematics* **2018**, *6*, 32. math6030032. [CrossRef]

24. Podlubny, I. *Fractional Differential Equations*; Academic Press: New York, NY, USA, 1999.

25. Gel'fand, I.M.; Shilov, G.E. *Generalized Functions*; Academic Press: New York, NY, USA, 1964; Volume I.

26. Li, C. Several results of fractional derivatives in $\mathcal{D}'(R^+)$. *Fract. Calc. Appl. Anal.* **2015**, *18*, 192–207. [CrossRef]

27. Matignon, D. Stability results for fractional differential equations with applications to control processing. In *Computational Engineering in System Applications*; IMACS, IEEE-SMC: Lille, France, 1996; Volume 2, pp. 963–968.

28. Babenko, Y.I. *Heat and Mass Transfer*; Khimiya: Leningrad, Russia, 1986. (In Russian)

29. Heaviside, O. On operators in physical mathematics, Part 1. *Proc. R. Soc.* **1893**, *52*, 504–529. [CrossRef]

30. Heaviside, O. On operators in physical mathematics, Part 2. *Proc. R. Soc.* **1893**, *54*, 105–143. [CrossRef]

31. Heaviside, O. *Electromagnetic Theory*; Ernest Benn Ltd.: London, UK, 1925; Volumes 1, 2 and 3.

32. Li, C. Several results on the commutative neutrix product of distributions. *Integral Transforms Spec. Funct.* **2007**, *18*, 559–568. [CrossRef]

33. Fisher, B.; Al-Sirehy, F. On the convolution and neutrix convolution of the functions $\sinh^{-1} x$ and x^r. *Sarajevo J. Math.* **2015**, *11*, 37–48. [CrossRef]

34. Fisher, B.; Ozcag, E. Some results on the neutrix composition of the delta function. *Filomat* **2012**, *26*, 1247–1256. [CrossRef]

35. Fisher, B.; Ozcag, E.; Al-Sirehy, F. On the composition and neutrix composition of the delta function and the function $\cosh^{-1}(|x|^{1/r} + 1)$. *Int. J. Anal. Appl.* **2017**, *13*, 161–169.

36. Fisher, B.; Taş, K. The convolution of functions and distributions. *J. Math. Anal. Appl.* **2005**, *306*, 364–374. [CrossRef]

37. Lazarova, L.; Jolevska-Tuneska, B.; Akturk, I.; Ozcag, E. Note on the distribution composition $(x_+^\mu)^\lambda$. *Bull. Malaysian Math. Soc.* **2016**. [CrossRef]

38. Ozcag, E.; Lazarova, L.; Jolevska-Tuneska, B. Defining compositions of $x_+^\mu, |x|^\mu, x^{-s}$ and $x^{-s} \ln |x|$ as neutrix limit of regular sequences. *Commun. Math. Stat.* **2016**, *4*, 63–80.

39. Li, C. The products on the unit sphere and even-dimension spaces. *J. Math. Anal. Appl.* **2005**, *305*, 97–106. [CrossRef]

40. Cheng, L.; Li, C. A commutative neutrix product of distributions on R^m. *Math. Nachr.* **1991**, *151*, 345–355.

41. Li, C. A review on the products of distributions. In *Mathematical Methods in Engineering*; Taş, K., Tenreiro Machado, J.A., Baleanu, D., Eds.; Springer: Dordrecht, The Netherlands, 2007; pp. 71–96.

![Σ] *mathematics*

![MDPI]

Article

F-Convex Contraction via Admissible Mapping and Related Fixed Point Theorems with an Application

Y. Mahendra Singh [1], Mohammad Saeed Khan [2] and Shin Min Kang [3,4,*]

[1] Department of Humanities and Basic Sciences, Manipur Institute of Technology, Takyelpat 795004, India;
ymahenmit@rediffmail.com

[2] Department of Mathematics and Statistics, Sultan Qaboos University, P.O. Box 36,
Al-Khod Muscat 123, Oman; mohammad@squ.edu.om

[3] Department of Mathematics and RINS, Gyeongsang National University, Jinju 52828, Korea

[4] Center for General Education, China Medical University, Taichung 40402, Taiwan

* Correspondence:smkang@gnu.ac.kr

Received: 8 May 2018; Accepted: 5 June 2018; Published: 20 June 2018

Abstract: In this paper, we introduce *F*-convex contraction via admissible mapping in the sense of Wardowski [Fixed points of a new type of contractive mappings in complete metric spaces. Fixed Point Theory Appl., 94 (2012), 6 pages] which extends convex contraction mapping of type-2 of Istrățescu [Some fixed point theorems for convex contraction mappings and convex non-expansive mappings (I), Libertas Mathematica, 1(1981), 151–163] and establish a fixed point theorem in the setting of metric space. Our result extends and generalizes some other similar results in the literature. As an application of our main result, we establish an existence theorem for the non-linear Fredholm integral equation and give a numerical example to validate the application of our obtained result.

Keywords: α-admissible mapping; α^*-admissible; *F*-contraction; α-*F*-convex contraction; fixed point; non-linear Fredholm integral equation

1. Introduction and Preliminaries

The Banach's contraction principle [1] first appeared in explicit form in 1922, where it was used to establish the existence of a solution for an integral equation. Since then, because of its simplicity, usefulness and constructiveness, it has become a very popular and a fundamental tool in solving existence problems arising not only in pure and applied mathematics but also in many branches of sciences, engineering, social sciences, economics and medical sciences. One of the most common generalizations of Banach's contraction is the 1971's Ćirić contraction [2] (also see [3]), in that he considered all possible six values $d(x,y), d(x,Tx), d(y,Ty), d(x,Ty), d(y,Tx)$ and $d(Tx,Ty)$ by combining x, y, Tx, Ty for all $x, y \in X$ and T (self mapping on a metric space (X,d)). Later, in 1982, Istrățescu [4] introduced the class of convex contractions in metric space, where he considered seven values $d(x,y), d(Tx,Ty), d(x,Tx), d(Tx,T^2x), d(y,Ty), d(Ty,T^2y)$ and $d(T^2x,T^2y)$ for all $x, y \in X$. Further, he showed with example (see Example 1.3, [4]) that T is in the class of convex contraction but it is not a contraction. Recently, some researchers studied on generalization of such class of mappings in the setting of various spaces (for example, Alghamdi et al. [5], Ghorbanian et al. [6], Latif et al. [7], Miandaragh et al. [8], Miculescu [9], etc.). Khan et al. [10], introduced the notion of generalized convex contraction mapping of type-2 by extending the generalized convex contraction (respectively, generalized convex contraction of order-2) of Miandaragh et al. [8] and the convex contraction mapping of type-2 of Istrățescu [4]. Very recently, Khan et al. [11], discussed the notions of (α, p)-convex contraction (respectively (α, p)-contraction) and asymptotically T^2-regular (respectively (T, T^2)-regular) sequence and, showed that (α, p)-convex contraction reduces to two-sided convex

contraction [4]. Further, they have also shown with examples that the notions of asymptotically
T-regular and T^2-regular sequences are independent to each other.

Generalizing the Banach contraction principle, Wardowski [12] introduced the notion of
F-contraction and proved a new fixed point theorem concerning F-contractions.

Definition 1. *[12] Let $F : R^+ \to R$ be a mapping satisfying the following:*

(F_1) *F is strictly increasing, i.e., for all $\alpha, \beta \in R^+$ such that $\alpha < \beta, F(\alpha) < F(\beta)$;*
(F_2) *For each sequence $\{\alpha_n\}_{n \in N}$ of positive numbers $\lim_{n \to \infty} \alpha_n = 0$ if and only if $\lim_{n \to \infty} F(\alpha_n) = -\infty$;*
(F_3) *There exists $k \in (0, 1)$ such that $\lim_{\alpha \to 0+} \alpha^k F(\alpha) = 0$.*

We denote $\mathcal{F}$, the set of all functions satisfying the above definition.

Definition 2. *[12] A mapping $T : X \to X$ is said to be an F-contraction on (X, d) if there exist $F \in \mathcal{F}$ and $\tau > 0$ such that $\forall x, y \in X$,*

$$d(Tx, Ty) > 0 \Rightarrow \tau + F(d(Tx, Ty)) \leq F(d(x, y)). \tag{1}$$

Example 1. *[12] The following functions $F : \mathbb{R}^+ \to \mathbb{R}$ are in $\mathcal{F}$.*

(i) *$F(\alpha) = \ln \alpha$;*
(ii) *$F(\alpha) = \ln \alpha + \alpha$;*
(iii) *$F(\alpha) = -\frac{1}{\sqrt{\alpha}}$;*
(iv) *$F(\alpha) = \ln(\alpha^2 + \alpha)$.*

By using F-contraction, Wardowski [12] proved a fixed point theorem which generalizes Banach's
contraction principle in a different way than in the known results from the literature.

Theorem 1. *[12] Let (X, d) be a complete metric space and $T : X \to X$ be an F-contraction. Then, we have*

(i) *T has a unique fixed point $z \in X$;*
(ii) *For all $x \in X$, the sequence $\{T^n x\}$ is convergent to $z \in X$.*

Definition 3. *([2,3]) Let (X, d) be a metric space and $T : X \to X$ be a mapping. Then, T is said to be orbitally continuous on X if $\lim_{i \to \infty} T^{n_i} x = z$ implies that $\lim_{i \to \infty} T^{n_i} x = Tz$.*

Let $T : X \to X$ be a mapping on a non-empty set X. We denote $Fix(T) = \{x : Tx = x$ for all $x \in X\}$.

Definition 4. *[13] Let $T : X \to X$ be a self mapping on a non-empty set X and $\alpha : X \times X \to [0, \infty)$ be a mapping, we say that T is an α-admissible if $x, y \in X$, $\alpha(x, y) \geq 1$ implies that $\alpha(Tx, Ty) \geq 1$.*

Obviously, $\alpha(.,.)$ may or may not be symmetric and $\alpha(x, y) = \alpha(y, x)$ (i.e., symmetric) if and only
if $x = y$.

Definition 5. *[14] Let $T : X \to X$ and $\alpha : X \times X \to (-\infty, +\infty)$. We say that T is said to be a triangular α-admissible if*

(T_1) *$\alpha(x, y)$ implies $\alpha(Tx, Ty) \geq 1, x, y \in X$;*
(T_2) *$\alpha(x, z) \geq 1$ and $\alpha(z, y) \geq 1$ imply $\alpha(x, y) \geq 1$ for all $x, y, z \in X$. (see for more examples Karapinar et al. [14]).*

Example 2. *Let* $X = [0, \infty)$. *Define* $T : X \to X$ *and* $\alpha : X \times X \to [0, \infty)$ *by* $Tx = \ln(1 + x)$ *for all* $x \in X$ *and*

$$\alpha(x, y) = \begin{cases} 1 + x, & \text{if } x \geq y; \\ 0, & \text{otherwise.} \end{cases}$$

Then, T is α-admissible as $\alpha(x, y) \geq 1$ implies that $\alpha(Tx, Ty) \geq 1$ for $x \geq y$ and $\alpha(x, y) = \alpha(y, x)$ for all $x = y$.

Definition 6. *[15] Let T be an α-admissible mapping on a non-empty set X. We say that X has the property (H) if for each $x, y \in Fix(T)$, there exists $z \in X$ such that $\alpha(x, z) \geq 1$ and $\alpha(y, z) \geq 1$.*

Definition 7. *An α-admissible mapping T is said to be an α^*-admissible, if for each $x, y \in Fix(T) \neq \varnothing$, we have $\alpha(x, y) \geq 1$. If $Fix(T) = \varnothing$, we say that T is vacuously α^*-admissible.*

The above definition is used by some authors without its nomenclature concerning the uniqueness of fixed point (for examples Alsulami et al. [16], Khan et al. [10], etc.).

Example 3. *Let* $X = [0, \infty)$. *Define* $T : X \to X$ *and* $\alpha : X \times X \to [0, \infty)$ *by* $Tx = 1 + x$ *for all* $x \in X$ *and*

$$\alpha(x, y) = \begin{cases} e^{2(x-y)}, & \text{if } x \geq y; \\ 0, & \text{otherwise.} \end{cases}$$

Then, T is α-admissible. Since T has no fixed point, i.e., $Fix(T) = \varnothing$, so T is vacuously α^*-admissible.

Example 4. *Let* $X = [0, \infty)$. *Define* $T : X \to X$ *and* $\alpha : X \times X \to [0, \infty)$ *by* $Tx = \frac{x^2}{2}$ *for all* $x \in X$ *and*

$$\alpha(x, y) = \begin{cases} 1, & \text{if } x, y \in [0, 2]; \\ 0, & \text{otherwise.} \end{cases}$$

Obviously, T is α-admissible and $Fix(T) = \{0, 2\}$. Then, T is α^*-admissible

Example 5. *Let* $X = [0, \infty)$. *Define* $T : X \to X$ *and* $\alpha : X \times X \to [0, \infty)$ *by* $Tx = \sqrt{\frac{x(x^2+2)}{3}}$ *for all* $x \in X$ *and*

$$\alpha(x, y) = \begin{cases} 1, & \text{if } x, y \in [0, 1]; \\ 0, & \text{otherwise.} \end{cases}$$

Obviously, T is α-admissible and $Fix(T) = \{0, 1, 2\}$. Note that T is not an α^*-admissible, since $\alpha(x, 2) = 0$ for $x \in \{0, 1\}$.

In the next section, we extend the notion of convex contraction [4] to an α-F-convex contraction and prove a fixed point theorem in the setting of metric space.

2. α-F-Convex Contraction

In this section, we discuss the class of α-F-convex contractions. Let T be a mapping on a metric space (X, d). We denote

$$M^p(x, y) = \max \{ d^p(x, y), d^p(Tx, Ty), d^p(x, Tx), d^p(Tx, T^2x), d^p(y, Ty), \\ d^p(Ty, T^2y) \}. \tag{2}$$

Definition 8. *A self mapping T on X is said to be an α-F-convex contraction, if there exist two functions* $\alpha : X \times X \to [0, \infty)$ *and* $F \in \mathcal{F}$ *such that*

$$d^p(T^2x, T^2y) > 0 \Rightarrow \tau + F\big(\alpha(x,y)d^p(T^2x, T^2y)\big) \leq F\big(M^p(x,y)\big) \tag{3}$$

for all $x, y \in X$ *where* $p \in [1, \infty)$ *and* $\tau > 0$.

Example 6. *Let* $F(\gamma) = \ln(\gamma)$, $\gamma > 0$. *Obviously,* $F \in \mathcal{F}$. *Let* $T : X \to X$ *be a self mapping on a metric space* (X, d). *Consider the convex contraction of type-2 (Istrăţescu [4]) taking with* $\alpha(x,y) = 1$ *for all* $x, y \in X$, $e^{-\tau} = k = \sum_{i=1}^{6} \alpha_i < 1$ *and* $\alpha_i \geq 0$ *for all* $i = 1, 2, ..., 6$.

$$d(T^2x, T^2y) \leq \alpha_1 d(x,y) + \alpha_2 d(Tx, Ty) + \alpha_3 d(x, Tx) + \alpha_4 d(Tx, T^2x)$$
$$+ \alpha_5 d(y, Ty) + \alpha_6 d(Ty, T^2y),$$

where $x, y \in X$ *with* $x \neq y$. *Then, we obtain*

$$\alpha(x,y)d(T^2x, T^2y) = d(T^2x, T^2y)$$
$$\leq \sum_{i=1}^{6} \alpha_i \max \big\{ d(x,y), d(Tx, Ty), d(x, Tx),$$
$$d(Tx, T^2x), d(y, Ty), d(Ty, T^2y) \big\},$$

which implies that

$$\alpha(x,y)d(T^2x, T^2y) \leq kM^1(x,y) = e^{-\tau}M^1(x,y).$$

Taking natural logarithm on both sides, we obtain

$$\tau + F\big(\alpha(x,y)d(T^2x, T^2y)\big) \leq F\big(M^1(x,y)\big).$$

Therefore,

$$d(T^2x, T^2y) > 0 \Rightarrow \tau + F\big(\alpha(x,y)d(T^2x, T^2y)\big) \leq F\big(M^1(x,y)\big)$$

for all $x, y \in X$. *This shows that T is an α-F-convex contraction with* $p = 1$.

Example 7. *Let* $X = [0,1]$ *with usual metric* $d(x,y) = |x - y|$. *Define a mapping* $T : X \to X$ *by* $Tx = \frac{x^2}{2} + \frac{1}{4}$ *for all* $x \in X$ *with* $\alpha(x,y) = 1$ *for all* $x, y \in X$. *Then, T is α-admissible. Further, we see that T is non-expansive, since we have*

$$|Tx - Ty| = \frac{1}{2}|x^2 - y^2| \leq |x - y| \quad \text{for all } x, y \in X.$$

Setting $F \in \mathcal{F}$ *such that* $F(\gamma) = \ln \gamma$, $\gamma > 0$. *Then, for all* $x, y \in X$ *with* $x \neq y$, *we obtain*

$$\alpha(x,y)|T^2x - T^2y| = |T^2x - T^2y|$$
$$= \frac{1}{8}\big(|(x^4 + x^2) - (y^4 + y^2)|\big)$$
$$\leq \frac{1}{8}\big(|x^4 - y^4| + |x^2 - y^2|\big)$$
$$\leq \frac{1}{2}|Tx - Ty| + \frac{1}{4}|x - y|$$
$$\leq \frac{3}{4}\max\big\{|Tx - Ty|, |x - y|\big\}$$
$$\leq e^{-\tau}M^1(x,y) \leq e^{-\tau}M^1(x,y),$$

where $-\tau = \ln\left(\frac{3}{4}\right)$. Taking natural logarithm on both sides, we obtain

$$\tau + F\big(\alpha(x,y)d(T^2x, T^2y)\big) \leq F\big(M^1(x,y)\big).$$

This shows that T is an α-F-convex contraction with $p = 1$.

Example 8. *Let $T : X \to X$, where $X = [0,1]$ with usual metric $d(x,y) = |x - y|$. Define $Tx = \frac{1-x^2}{2}$, for all $x \in X$ and $\alpha(x,y) = 1$ for all $x, y \in X$. Then, T is α-admissible. Setting $F \in \mathcal{F}$ such that $F(\gamma) = \ln\gamma$, $\gamma > 0$. Then, for all $x, y \in X$ with $x \neq y$, we obtain*

$$|Tx - Ty| = \frac{1}{2}|x^2 - y^2| \leq |x - y|$$

and

$$\begin{aligned}
\alpha(x,y)|T^2x - T^2y| &= \frac{1}{8}\big(|(2x^2 - x^4) - (2y^2 - y^4)|\big)\\
&= \frac{1}{8}\big(2|x^2 - y^2| + |x^4 - y^4|\big)\\
&\leq \frac{1}{2}|x - y| + \frac{1}{4}|x^2 - y^2|\\
&= \frac{1}{2}|x - y| + \frac{1}{2}|Tx - Ty|\\
&\leq \max\{|x - y|, |Tx - Ty|\}\\
&\leq M^1(x,y).
\end{aligned}$$

Taking natural logarithm on both sides, we obtain

$$F\big(\alpha(x,y)d(T^2x, T^2y)\big) \leq F\big(M^1(x,y)\big).$$

However, there does not exist $\tau > 0$ such that

$$\tau + F\big(\alpha(x,y)d(T^2x, T^2y)\big) \leq F\big(M^1(x,y)\big).$$

Therefore, T is not an α-F-convex contraction with $p = 1$. Also, consider

$$|Tx - Ty|^2 = \frac{1}{4}|x^2 - y^2|^2 \leq |x - y|^2$$

and

$$\begin{aligned}
\alpha(x,y)|T^2x - T^2y|^2 &= \frac{1}{64}|(2x^2 - x^4) - (2y^2 - y^4)|^2\\
&\leq \frac{1}{64}\big(4|x^2 - y^2|^2 + |x^4 - y^4|^2\big)\\
&= \frac{1}{16}|x^2 - y^2|^2 + \frac{1}{64}|x^4 - y^4|^2\\
&\leq \frac{1}{4}|x - y|^2 + \frac{1}{16}|x^2 - y^2|^2\\
&\leq \frac{5}{16}\max\{|x - y|^2, |Tx - Ty|^2\}\\
&= \frac{5}{16}M^2(x,y) = e^{-\tau}M^2(x,y).
\end{aligned}$$

Taking natural logarithm on both sides, we obtain

$$\tau + F\big(\alpha(x,y)d(T^2x, T^2y)\big) \le F\big(M^2(x,y)\big),$$

where $-\tau = \ln \frac{5}{16}$. *Therefore, T is an* α-*F-convex contraction with* $p = 2$.

3. Fixed Point Results of an α-*F*-Contraction

We prove the following lemma which will be used in the sequel.

Lemma 1. *Let* (X, d) *be a metric space and* $T : X \to X$ *be an* α-*F-convex contraction satisfying the following conditions:*

(i)   *T is* α-*admissible;*
(ii)   *there exists* $x_0 \in X$ *such that* $\alpha(x_0, Tx_0) \ge 1$.

    Define a sequence $\{x_n\}$ *in* X *by* $x_{n+1} = Tx_n = T^{n+1}x_0$ *for all* $n \ge 0$. *Then* $\{d^p(x_n, x_{n+1})\}$ *is strictly non-increasing sequence in* X.

Proof. Let $x_0 \in X$ be such that $\alpha(Tx_0, x_0) \ge 1$ and $\{x_n\}$ is a sequence defined by $x_{n+1} = Tx_n$ for all $n \in \mathbb{N} \cup \{0\}$. Since T is α-admissible, $\alpha(x_0, x_1) = \alpha(x_0, Tx_0) \ge 1$ implies that $\alpha(x_2, x_3) = \alpha(Tx_1, T^2x_0) \ge 1$. One can obtain inductively that $\alpha(x_n, x_{n+1}) \ge 1$ for all $n \ge 0$. Assume that $x_n \ne x_{n+1}$ for all $n \ge 0$. Then, $d(x_n, x_{n+1}) > 0$ for all $n \ge 0$. Setting $v = \max\{d^p(x_0, x_1), d^p(x_1, x_2)\}$. From (2), taking $x = x_0$ and $y = x_1$, we obtain

$$\begin{aligned}
M^p(x_0, x_1) &= \max\big\{d^p(x_0, x_1), d^p(Tx_0, Tx_1), d^p(x_0, Tx_0), \\
&\qquad\qquad d^p(Tx_0, T^2x_0), d^p(x_1, Tx_1), d^p(Tx_1, T^2x_1)\big\} \\
&= \max\big\{d^p(x_0, x_1), d^p(x_1, x_2), d^p(x_0, x_1), \\
&\qquad\qquad d^p(x_1, x_2), d^p(x_1, x_2), d^p(x_2, x_3)\big\} \\
&= \max\big\{d^p(x_0, x_1), d^p(x_1, x_2), d^p(x_2, x_3)\big\}.
\end{aligned} \tag{4}$$

Since F is strictly increasing and $\alpha(x_0, x_1) \ge 1$, by (3) and (4), we obtain

$$\begin{aligned}
F\big(d^p(x_2, x_3)\big) &= F\big(d^p(T^2x_0, T^2x_1)\big) \\
&\le F\big(\alpha(x_0, x_1)d^p(T^2x_0, T^2x_1)\big) \\
&\le F\big(M^p(x_0, x_1)\big) - \tau \\
&= F\big(\max\{d^p(x_0, x_1), d^p(x_1, x_2), d^p(x_2, x_3)\}\big) - \tau \\
&\le F\big(\max\{v, d^p(x_2, x_3)\}\big) - \tau.
\end{aligned} \tag{5}$$

If $\max\{v, d^p(x_2, x_3)\} = d^p(x_2, x_3)$, then (5) gives

$$F\big(d^p(x_2, x_3)\big) \le F\big(d^p(x_2, x_3)\big) - \tau < F\big(d^p(x_2, x_3)\big).$$

This is a contradiction. It follows that

$$F\big(d^p(x_2, x_3)\big) \le F(v) - \tau < F(v).$$

Since $\tau > 0$ and F is strictly increasing, it follows that

$$d^p(x_2, x_3) < v = \max\big\{d^p(x_0, x_1), d^p(x_1, x_2)\big\}.$$

Again, from (2) taking with $x = x_1$ and $y = x_2$, we obtain

$$
\begin{aligned}
M^p(x_1, x_2) &= \max \big\{ d^p(x_1, x_2), d^p(Tx_1, Tx_2), d^p(x_1, Tx_1), \\
&\qquad\quad d^p(Tx_1, T^2x_1), d^p(x_2, Tx_2), d^p(Tx_2, T^2x_2) \big\} \\
&= \max \big\{ d^p(x_1, x_2), d^p(x_2, x_3), d^p(x_1, x_2), \\
&\qquad\quad d^p(x_2, x_3), d^p(x_2, x_3), d^p(x_3, x_4) \big\} \\
&= \max \big\{ d^p(x_1, x_2), d^p(x_2, x_3), d^p(x_3, x_4) \big\}.
\end{aligned}
\tag{6}
$$

By (3) and (6), we obtain

$$
\begin{aligned}
F\big(d^p(x_3, x_4)\big) &= F\big(d^p(T^2x_1, T^2x_2)\big) \\
&\leq F\big(\alpha(x_1, x_2) d^p(T^2x_1, T^2x_2)\big) \\
&\leq F\big(M^p(x_1, x_2)\big) - \tau \\
&= F\big(\max\{d^p(x_1, x_2), d^p(x_2, x_3), d^p(x_3, x_4)\}\big) - \tau.
\end{aligned}
$$

If $\max \big\{ d^p(x_1, x_2), d^p(x_2, x_3), d^p(x_3, x_4) \big\} = d^p(x_3, x_4)$, then we obtain

$$
F\big(d^p(x_3, x_4)\big) \leq F\big(d^p(x_3, x_4)\big) - \tau < F\big(d^p(x_3, x_4)\big).
$$

This is again a contradiction. It follows that

$$
\max\{d^p(x_1, x_2), d^p(x_2, x_3)\} > d^p(x_3, x_4).
$$

Therefore, $v > d^p(x_2, x_3) > d^p(x_3, x_4)$. Continuing in this process, one can prove inductively that $\{d^p(x_n, x_{n+1})\}$ is a strictly non-increasing sequence in X. $\quad\square$

Theorem 2. *Let (X, d) be a complete metric space and $T : X \to X$ be an α-F-convex contraction satisfying the following conditions:*

(i) T is α-admissible;
(ii) there exists $x_0 \in X$ such that $\alpha(x_0, Tx_0) \geq 1$;
(iii) T is continuous or, orbitally continuous on X.

Then, T has a fixed point in X. Further, if T is α^*-admissible, then T has a unique fixed point $z \in X$. Moreover, for any $x_0 \in X$ if $x_{n+1} = T^{n+1}x_0 \neq Tx_n$ for all $n \in \mathbb{N} \cup \{0\}$, then $\lim_{n \to \infty} T^n x_0 = z$.

Proof. Let $x_0 \in X$ be such that $\alpha(Tx_0, x_0) \geq 1$ and define a sequence $\{x_n\}$ by $x_{n+1} = Tx_n$ for all $n \in \mathbb{N} \cup \{0\}$. If $x_{n_0} = x_{n_0+1}$, i.e., $Tx_{n_0} = x_{n_0}$ for some $n_0 \in \mathbb{N} \cup \{0\}$, then x_{n_0} is a fixed point of T.

Now, we assume that $x_n \neq x_{n+1}$ for all $n \geq 0$. Then, $d(x_n, x_{n+1}) > 0$ for all $n \geq 0$. Since T is α-admissible, $\alpha(x_0, Tx_0) \geq 1$ implies that $\alpha(x_1, x_2) = \alpha(Tx_0, T^2x_0) \geq 1$. Therefore, one can obtain inductively that $\alpha(x_n, x_{n+1}) = \alpha(T^n x_0, T^{n+1}x_0) \geq 1$ for all $n \geq 0$. Setting $v = \max\{d^p(x_0, x_1), d^p(x_1, x_2)\}$.

Now from (2), taking $x = x_{n-2}$ and $y = x_{n-1}$, where $n \geq 2$, we obtain

$$
\begin{aligned}
M^p(x_{n-2}, x_{n-1}) &= \max \big\{ d^p(x_{n-2}, x_{n-1}), d^p(Tx_{n-2}, Tx_{n-1}), d^p(x_{n-2}, Tx_{n-2}), \\
&\qquad\quad d^p(Tx_{n-2}, T^2x_{n-2}), d^p(x_{n-1}, Tx_{n-1}), d^p(Tx_{n-1}, T^2x_{n-1}) \big\} \\
&= \max \big\{ d(x_{n-2}, x_{n-1}), d(x_{n-1}, x_n), d(x_{n-2}, x_{n-1}), \\
&\qquad\quad d^p(x_{n-1}, x_n), d^p(x_{n-1}, x_n), d^p(x_n, x_{n+1}) \big\} \\
&= \max \big\{ d^p(x_{n-2}, x_{n-1}), d^p(x_{n-1}, x_n), d^p(x_n, x_{n+1}) \big\}.
\end{aligned}
$$

Since F is strictly increasing and T is α-admissible, by (3), we obtain

$$
\begin{aligned}
F\big(d^p(x_n, x_{n+1})\big) &= F\big(d^p(T^2 x_{n-2}, T^2 x_{n-1})\big) \\
&\leq F\big(\alpha(x_{n-2}, x_{n-1}) d^p(T^2 x_{n-2}, T^2 x_{n-1})\big) \\
&\leq F\big(M^p(x_{n-2}, x_{n-1})\big) - \tau \\
&\leq F\big(\max\big\{d^p(x_{n-2}, x_{n-1}), d^p(x_{n-1}, x_n), d^p(x_n, x_{n+1})\big\}\big) - \tau.
\end{aligned}
$$

If $\max\big\{d^p(x_{n-2}, x_{n-1}), d^p(x_{n-1}, x_n), d^p(x_n, x_{n+1})\big\} = d^p(x_n, x_{n+1})$, then we obtain

$$
F\big(d^p(x_n, x_{n+1})\big) \leq F\big(d^p(x_n, x_{n+1})\big) - \tau < F\big(d^p(x_n, x_{n+1})\big).
$$

This is a contradiction. Therefore,

$$
F\big(d^p(x_n, x_{n+1})\big) \leq F\big(\max\big\{d^p(x_{n-2}, x_{n-1}), d^p(x_{n-1}, x_n)\big\}\big) - \tau.
$$

By Lemma 1, $\{d^p(x_n, x_{n+1})\}$ is strictly non-increasing sequence. Therefore,

$$
F\big(d^p(x_n, x_{n+1})\big) \leq F\big(d^p(x_{n-2}, x_{n-1})\big) - \tau \leq \cdots \leq F(v) - l\tau, \tag{7}
$$

whenever $n = 2l$ or $n = 2l + 1$ for $l \geq 1$.

From (6), we obtain

$$
\lim_{n \to \infty} F\big(d^p(x_n, x_{n+1})\big) = -\infty. \tag{8}
$$

Therefore, by (F_2) with (8), we obtain

$$
\lim_{n \to \infty} d(x_n, x_{n+1}) = 0. \tag{9}
$$

From (F_3), there exists $k \in (0, 1)$ such that

$$
\lim_{n \to \infty} \big[d^p(x_n, x_{n+1})\big]^k F\big(d^p(x_n, x_{n+1})\big) = 0. \tag{10}
$$

Also, from (7), we obtain

$$
\big[d^p(x_n, x_{n+1})\big]^k \big[F\big(d^p(x_n, x_{n+1})\big) - F(v)\big] \leq -\big[d^p(x_n, x_{n+1})\big]^k l\tau \leq 0, \tag{11}
$$

where $n = 2l$ or $n = 2l + 1$ for $l \geq 1$.

Letting $n \to \infty$ in (11) together with (9) and (10), we obtain

$$
\lim_{n \to \infty} l\big[d(x_n, x_{n+1})\big]^k = 0. \tag{12}
$$

Now, it arises in the following cases.

Case-(i): If n is even and $n \geq 2$, then from (12), we obtain

$$
\lim_{n \to \infty} n\big[d(x_n, x_{n+1})\big]^k = 0. \tag{13}
$$

Case-(ii): If n is odd and $n \geq 3$, then from (12), we obtain

$$
\lim_{n \to \infty} (n-1)\big[d(x_n, x_{n+1})\big]^k = 0. \tag{14}
$$

Using (9), (14) gives

$$
\lim_{n \to \infty} n\big[d(x_n, x_{n+1})\big]^k = 0. \tag{15}
$$

It may be observed from the above cases that, there exists $n_1 \in \mathbb{N}$ such that

$$n\big[d(x_n, x_{n+1})\big]^k \leq 1 \quad \text{for all } n \geq n_1.$$

Therefore, we obtain

$$d(x_n, x_{n+1}) \leq \frac{1}{n^{\frac{1}{k}}} \quad \text{for all } n \geq n_1.$$

Now, we show that $\{x_n\}$ is a Cauchy sequence. For all $p > q \geq n_1$, we obtain

$$d(x_p, x_q) \leq d(x_p, x_{p-1}) + d(x_{p-1}, x_{p-2}) + \cdots + d(x_{q+1}, x_q)$$
$$< \sum_{i=q}^{\infty} d(x_i, x_{i+1}) \leq \sum_{i=q}^{\infty} \frac{1}{i^{\frac{1}{k}}}.$$

Since $\sum_{i=q}^{\infty} \frac{1}{i^{\frac{1}{k}}}$ is convergent, taking $q \to \infty$, we get $\lim_{p,q\to\infty} d(x_p, x_q) = 0$. This shows that $\{x_n\}$ is a Cauchy sequence in X. Since X is complete, there exists $z \in X$ such that $\lim_{n\to\infty} x_n = z$. Now we prove that z is a fixed point of T. Suppose T is continuous, then

$$d(z, Tz) = \lim_{n\to\infty} d(x_n, Tx_n) = \lim_{n\to\infty} d(x_n, x_{n+1}) = 0.$$

This shows that z is a fixed point of T.

Again, we suppose that T is orbitally continuous on X, then $x_{n+1} = Tx_n = T(T^n x_0) \to Tz$ as $n \to \infty$. Since (X, d) is complete this implies that $Tz = z$. Therefore, $Fix(T) \neq \emptyset$.

Further, we suppose that T is α^*-admissible, it follows that for all $z, z^* \in Fix(T)$, we have $\alpha(z, z^*) \geq 1$. From (2) and (3), we obtain

$$\begin{aligned}
F\big(d^p(z, z^*)\big) &= F\big(d^p(T^2z, T^2z^*)\big) \\
&= F\big(\alpha(z, z^*)d^p(T^2z, T^2z^*)\big) \\
&\leq F\big(M^p(z, z^*)\big) - \tau \\
&= F\big(\max\{d^p(z, z^*), d^p(Tz, Tz^*), d^p(z, Tz), d^p(Tz, T^2z), \\
&\quad d^p(z^*, Tz^*), d^p(Tz^*, T^2z^*)\}\big) - \tau \\
&= F\big(d^p(z, z^*)\big) - \tau.
\end{aligned}$$

Since $\tau > 0$ and using F is strictly increasing, we obtain

$$F\big(d(z, z^*)\big) < F\big(d(z, z^*)\big).$$

This is a contradiction. Therefore, T has a unique fixed point in X. $\square$

One can verify the validity of Theorem 2 with Examples 7 and 8 with proper choose of $\alpha(x, y)$ and p.

Corollary 1. *Let (X, d) be a complete metric space and $\alpha : X \times X \to [0, \infty)$ be a function. Suppose that $T : X \to X$ be a self mapping satisfying the following conditions:*

(i) for all $x, y \in X$

$$\alpha(x, y)d(T^2x, T^2y) \leq k \max\{d(x, y), d(Tx, Ty), d(x, Tx), d(Tx, T^2x), \\ d(y, Ty), d(Ty, T^2y)\} \tag{16}$$

where $k \in [0, 1)$;

(ii) T *is α-admissible;*

(iii) there exists $x_0 \in X$ such that $\alpha(x_0, Tx_0) \geq 1$;

(iv) T is continuous or, orbitally continuous on X.

Then, T has a fixed point in X. Further, if T is an α^*-admissible, then T has a unique fixed point $z \in X$. Moreover, for any $x_0 \in X$ if $x_{n+1} = T^{n+1}x_0 \neq T^n x_0$ for all $n \in \mathbb{N} \cup \{0\}$, then $\lim_{n \to \infty} T^n x_0 = z$.

Proof. Setting $F(\gamma) = \ln(\gamma), \gamma > 0$. Obviously, $F \in \mathcal{F}$. Taking natural logarithm on both sides of (16), we obtain

$$- \ln k + \ln \alpha(x,y) d(T^2 x, T^2 y)$$
$$\leq \ln \left(\max \left\{ d(x,y), d(Tx, Ty), d(x, Tx), d(Tx, T^2 x), d(y, Ty), d(Ty, T^2 y) \right\} \right),$$

which implies that

$$\tau + F\left(\alpha(x,y) d(T^2 x, T^2 y) \right) \leq F\left(M^1(x,y) \right)$$

for all $x, y \in X$ with $x \neq y$ where $\tau = - \ln k$. This shows that T is α-F-convex contraction with $p = 1$. Thus, all the conditions of Theorem 2 are satisfied and hence, T has a unique fixed point in X. $\square$

The following is proved in Khan et al. ([10], [Theorem 2.2]) by extending convex contraction of type-2 (Istrăţescu [4]) to generalized convex contraction of type-2 in the setting of b-metric space .

Corollary 2. *Let (X, d) be a complete metric space and $\alpha : X \times X \to [0, \infty)$ be a function. Suppose that $T : X \to X$ be a self mapping satisfying the following conditions:*

(i)   for all $x, y \in X$

$$\alpha(x,y) d(T^2 x, T^2 y) \leq \alpha_1 d(x,y) + \alpha_2 d(Tx, Ty) + \alpha_3 d(x, Tx) + \alpha_4 d(Tx, T^2 x)$$
$$+ \alpha_5 d(y, Ty) + \alpha_6 d(Ty, T^2 y) \}$$

where $0 \leq \alpha_i < 1, i = 1, 2, ..., 6$ such that $\sum_{i=1}^{6} \alpha_i < 1$;

(ii) T is α-admissible;

(iii) there exists $x_0 \in X$ such that $\alpha(x_0, Tx_0) \geq 1$;

(iv) T is continuous or, orbitally continuous on X.

Then, T has a fixed point in X. Further, if T is α^*-admissible, then T has a unique fixed point $z \in X$. Moreover, for any $x_0 \in X$ if $x_{n+1} = T^{n+1}x_0 \neq T^n x_0$ for all $n \in \mathbb{N} \cup \{0\}$, then $\lim_{n \to \infty} T^n x_0 = z$.

Proof. Setting $F(\gamma) = \ln(\gamma), \gamma > 0$. Obviously, $F \in \mathcal{F}$. For all $x, y \in X$ with $x \neq y$, we obtain

$$\alpha(x,y) d(T^2 x, T^2 y) = d(T^2 x, T^2 y)$$
$$\leq \alpha_1 d(x,y) + \alpha_2 d(Tx, Ty) + \alpha_3 d(x, Tx) + \alpha_4 d(Tx, T^2 x)$$
$$+ \alpha_5 d(y, Ty) + \alpha_6 d(Ty, T^2 y) \}$$
$$\leq k \max \left\{ d(x,y), d(Tx, Ty), d(x, Tx), d(Tx, T^2 x), \right.$$
$$\left. d(y, Ty), d(Ty, T^2 y) \right\},$$

where $k = \sum_{i=1}^{6} \alpha_i < 1$. Therefore, by above Corollary 1, T has a unique fixed point in X. $\square$

Corollary 3. *Let T be a continuous mapping on a complete metric space* (X, d) *into itself. If there exists* $k \in [0, 1)$ *satisfying the following inequality*

$$d(T^2 x, T^2 y) \leq k \max \{ d(x, y), d(Tx, Ty), d(x, Tx), d(Tx, T^2 x),$$
$$d(y, Ty), d(Ty, T^2 y) \}$$

for all $x, y \in X$, *then T has a unique fixed point in X.*

4. Application

In this section, we apply our result to establish an existence theorem for the non-linear Fredholm integral equation and give a numerical example to validate the application of our obtained result.

Let $X = C[a, b]$ be a set of all real continuous functions on $[a, b]$ equipped with metric $d(f, g) = |f - g| = \max_{t \in [a,b]} |f(t) - g(t)|$ for all $f, g \in C[a, b]$. Then, (X, d) is a complete metric space. Now, we consider the non-linear Fredholm integral equation

$$x(t) = v(t) + \frac{1}{b - a} \int_a^b K(t, s, x(s)) ds, \tag{17}$$

where $t, s \in [a, b]$. Assume that $K : [a, b] \times [a, b] \times X \to \mathbb{R}$ and $v : [a, b] \to \mathbb{R}$ continuous, where $v(t)$ is a given function in X.

Theorem 3. *Suppose that* (X, d) *is a metric space equipped with metric* $d(f, g) = |f - g| = \max_{t \in [a,b]} |f(t) - g(t)|$ *for all* $f, g \in X$ *and* $T : X \to X$ *be a continuous operator on X defined by*

$$Tx(t) = v(t) + \frac{1}{b - a} \int_a^b K(t, s, x(s)) ds. \tag{18}$$

If there exists $k \in [0, 1)$ *such that for all* $x, y \in X$ *with* $x \neq y$ *and* $s, t \in [a, b]$ *satisfying the following inequality*

$$|K(t, s, Tx(s)) - K(t, s, Ty(s))| \leq k \max \{ |x(s) - y(s)|, |Tx(s) - Ty(s)|,$$
$$|x(s) - Tx(s)|, |Tx(s) - T^2 x(s)|, \tag{19}$$
$$|y(s) - Ty(s)|, |Ty(s) - T^2 y(s)| \},$$

then the integral operator defined by (18) *has a unique solution* $z \in X$ *and for each* $x_0 \in X$, $Tx_n \neq x_n$ *for all* $n \in \mathbb{N} \cup \{0\}$, *we have* $\lim_{n \to \infty} Tx_n = z$.

Proof. We define $\alpha : X \times X \to [0, \infty)$ such that $\alpha(x, y) = 1$ for all $x, y \in X$. Therefore, T is α-admissible. Setting $F \in \mathcal{F}$ such that $F(\gamma) = \ln(\gamma)$, $\gamma > 0$. Let $x_0 \in X$ and define a sequence $\{x_n\}$ in X by $x_{n+1} = Tx_n = T^{n+1} x_0$ for all $n \geq 0$. By (18), we obtain

$$x_{n+1} = Tx_n(t) = v(t) + \frac{1}{b - a} \int_a^b K(t, s, x_n(s)) ds. \tag{20}$$

We show that T is α-F-convex contraction on $C[a,b]$. Using (18) and (19), we obtain

$$|T^2x(t) - T^2y(t)| = \frac{1}{|b-a|}\left|\int_a^b K(t,s,Tx(s))ds - \int_a^b K(t,s,Ty(s))ds\right|$$

$$\leq \frac{1}{|b-a|}\int_a^b |K(t,s,Tx(s)) - K(t,s,Ty(s))|ds$$

$$\leq \frac{k}{|b-a|}\int_a^b \max\{|x(s)-y(s)|, |Tx(s)-Ty(s)|, |x(s)-Tx(s)|,$$
$$|Tx(s)-T^2x(s)|, |y(s)-Ty(s)|, |Ty(s)-T^2y(s)|\}ds.$$

Taking maximum on both sides for all $t \in [a,b]$, we obtain

$$d(T^2x, T^2y) = \max_{t\in[0,1]} |T^2x(t) - T^2y(t)|$$

$$\leq \frac{k}{|b-a|}\max_{t\in[a,b]}\int_a^b \max\{|x(s)-y(s)|, |Tx(s)-Ty(s)|, |x(s)-Tx(s)|,$$
$$|Tx(s)-T^2x(s)|, |y(s)-Ty(s)|, |Ty(s)-T^2y(s)|\}ds$$

$$\leq \frac{k}{|b-a|}\max\Big[\max_{r\in[a,b]}\{|x(r)-y(r)|, |Tx(r)-Ty(r)|, |x(r)-Tx(r)|,$$
$$|Tx(r)-T^2x(r)|, |y(r)-Ty(r)|, |Ty(r)-T^2y(r)|\}\Big]\int_a^b ds$$

$$= k\max\{d(x,y), d(Tx,Ty), d(x,Tx), d(Tx,T^2x), d(y,Ty), d(Ty,T^2y)\}$$

$$= kM^1(x,y).$$

Therefore

$$\alpha(x,y)d(T^2x, T^2y) \leq kM^1(x,y).$$

Now, taking natural logarithm on both sides, we obtain

$$-\ln k + \ln[\alpha(x,y)d(T^2x,T^2y)] \leq \ln M^1(x,y).$$

Thus, we obtain

$$\tau + F(\alpha(x,y)d(T^2x,T^2y)) \leq F(M^1(x,y)),$$

where $-\ln k = \tau$. This shows that T is α-F-convex contraction with $p=1$ for all $x,y \in X$ with $x \neq y$. Since T is α-admissible and $X = C[a,b]$ is complete metric space. Therefore, the iteration scheme (4.4) converges to some point $z \in X$, i.e., $\lim_{n\to\infty} x_n \to z$. By continuity of T, one can prove that T has a fixed point, i.e., $Tz = z$. Consequently, $Fix(T) \neq \emptyset$. Also, for all $x,y \in Fix(T)$, $\alpha(z,z^*) = 1$ follows that T is α^*-admissible. Thus, all the conditions of Theorem 2 are satisfied and hence, the integral operator T defined by (18) has a unique solution $z \in X$. $\square$

The following example shows the existence of unique solution of an integral operator satisfying all the hypothesis in Theorem 3, however, one can also check that the following example does not satisfy F-contraction.

Example 9. *Let $X = C[0,1]$ be a set of all continuous functions defined on $[0,1]$ equipped with metric $d(f,g) = |f-g| = \max_{t\in[0,1]} |f(t) - g(t)|$ for all $f,g \in X$. Let $T : X \to X$ be the operator defined by*

$$Tx(t) = v(t) + \int_0^1 K(t,s,x(s))ds. \tag{21}$$

Therefore,

$$T^2x(t) = v(t) + \int_0^1 K(t, s, Tx(s))ds$$

$$= v(t) + \int_0^1 K(t, s, v(t) + \int_0^1 K(t, s, x(t))ds)ds. \tag{22}$$

Letting $v(t) = \frac{8}{15}t^2$ *and* $K(t, s, x(s)) = \frac{1}{4}t^2(1 + s)(x^2(s) + 1)$. *Then,* (21) *becomes*

$$Tx(t) = \frac{8}{15}t^2 + \int_0^1 \frac{1}{4}t^2(1 + s)(x^2(s) + 1)ds. \tag{23}$$

Then, (i) $v(t)$ *and* $K(t, s, x(s))$ *are continuous,*
(ii) $\max \frac{1}{4}|t^2(1 + s)| \leq \frac{1}{2}$ *for all* $(t, s) \in [0, 1] \times [0, 1]$,
(iii) $Tx \in X$ *for all* $x \in X$,
(iv) For all $x, y \in X$ *with* $x \neq y$ *and* $(t, s) \in [0, 1] \times [0, 1]$ *and using* (21) *and* (22)*, we obtain*

$$|Tx(t) - Ty(t)| = \left| \int_0^1 K(t, s, x(s))ds - \int_0^1 K(t, s, y(s))ds \right|$$

$$\leq \int_0^1 \left| K(t, s, x(s)) - K(t, s, y(s))ds \right| ds$$

$$= \int_0^1 \left| \frac{t^2(1 + s)}{4}(x^2(s) - y^2(s)) \right| ds$$

Taking maximum on both sides for all $t \in [a, b]$*, we obtain*

$$|Tx - Ty| = \max_{t \in [0,1]} |Tx(t) - Ty(t)|$$

$$\leq \max_{t \in [0,1]} \int_0^1 \left| \frac{t^2(1 + s)}{4}(x^2(s) - y^2(s)) \right| ds$$

$$\leq \frac{1}{2} \max_{\rho \in [0,1]} \left| x^2(\rho) - y^2(\rho) \right| \int_0^1 ds$$

$$\leq |x - y|$$

From the above, one can verify that the integral operator T is not an F-contraction. Also, we have

$$|T^2x(t) - T^2y(t)| = \left| \int_0^1 K(t, s, Tx(s))ds - \int_0^1 K(t, s, Ty(s))ds \right|$$

$$= \left| \int_0^1 K\left(t, s, v(t) + \int_0^1 K(t, s, x(t))ds\right)ds \right.$$

$$\left. - \int_0^1 K\left(t, s, v(t) + \int_0^1 K(t, s, y(t))ds\right)ds \right|$$

$$\leq \int_0^1 \left| K\left(t, s, v(t) + \int_0^1 K(t, s, x(t))ds\right)ds \right.$$

$$\left. - K\left(t, s, v(t) + \int_0^1 K(t, s, y(t))ds\right) \right| ds$$

$$= \int_0^1 \left| \frac{t^2(1 + s)}{4} \int_0^1 \left[\frac{t^2(1 + s)}{8}(x^2(s) - y^2(s)) \right] ds \right| ds$$

$$\leq \int_0^1 \int_0^1 \left| \left[\frac{t^2(1 + s)}{4} \right]^2 (x^2(s) - y^2(s)) \right| ds ds.$$

Taking maximum on both sides for all $t \in [a, b]$, we obtain

$$|T^2x - T^2y| = \max_{t \in [0,1]} |T^2x(t) - T^2y(t)|$$

$$\leq \max_{t \in [0,1]} \int_0^1 \int_0^1 \left| \left[\frac{t^2(1+s)}{4} \right]^2 (x^2(s) - y^2(s)) \right| ds\,ds$$

$$\leq \frac{1}{4} \max_{\rho \in [0,1]} |x^2(\rho) - y^2(\rho)| \int_0^1 \int_0^1 ds\,ds$$

$$= \frac{1}{4}|x^2 - y^2| \leq \frac{1}{2}|x - y| \leq \frac{1}{2}M^1(x,y)$$

$$= e^{-\tau} M^1(x,y),$$

where $\ln \frac{1}{2} = -\tau$. Setting $\alpha : X \times X \to [0, \infty)$ by $\alpha(x,y) = 1$ for all $x, y \in X$ and $F \in \mathcal{F}$ such that $F(\gamma) = \ln(\gamma), \gamma > 0$. Therefore, we obtain

$$\alpha(x,y)|T^2x - T^2y| \leq e^{-\tau} M^1(x,y).$$

Taking natural logarithm on both sides, we obtain

$$\tau + \ln \alpha(x,y)d(T^2x, T^2y) \leq \ln M^1(x,y),$$

that is,

$$\tau + F\big(\alpha(x,y)d(T^2x, T^2y)\big) \leq F\big(M^1(x,y)\big).$$

This shows that T is α-F-convex contraction with $p = 1$ for all $x, y \in X$. Thus, all the conditions of Theorem 3 are satisfied and therefore, the integral Equation (21) has a unique solution. One can verify that $x(t) = t^2$ is the exact solution of the equation (21). Using the iteration scheme (20), (23) becomes

$$x_{n+1} = Tx_n(t) = \frac{8}{15}t^2 + \frac{1}{4}\int_0^1 \left[t^2(1+s)\{x_n^2(s) + 1\} \right] ds. \tag{24}$$

Letting $x_0(t) = 0$ be an initial solution. Putting $n = 0, 1, 2, ...,$ successively in (24), we obtain

$$x_1(t) = 0.90833333t^2, \quad x_2(t) = 0.98396470t^2, \quad x_3(t) = 0.99708377t^2$$
$$x_4(t) = 0.99946614t^2, \quad x_5(t) = 0.99990215t^2, \quad x_6(t) = 0.99998206t^2$$
$$x_7(t) = 0.99999671t^2, \quad x_8(t) = 0.99999940t^2, \quad x_9(t) = 0.99999985t^2$$
$$x_{10}(t) = 0.99999998t^2, \quad x_{11}(t) = t^2.$$

Therefore, $x(t) = t^2$ is the unique solution.

Author Contributions: All authors contributed equally in writing this article. All authors read and approved the final manuscript.

Acknowledgments: The authors are thankful to the anonymous referee for their valuable constructive comments and suggestions to improve the quality of the manuscript.

Conflicts of Interest: The authors declare no conflict of interest.

References

1. Banach, S. Sur les opérations dans les ensembles abstraits et leur application aux équations intégrales. *Fund. Math.* **1922**, *3*, 133–181. [CrossRef]
2. Ćirić, L.B. Generalized contractions and fixed point theorems. *Publ. Inst. Math.* **1971**, *12*, 19–26.
3. Ćirić, L.B. A generalizatin of Banach's contraction principle. *Proc. Am. Math. Soc.* **1974**, *45*, 267–273. [CrossRef]

4. Istrățescu, V.I. Some fixed point theorems for convex contraction mappings and convex nonexpansive mappings (I). *Liberta Math.* **1981**, *1*, 151–163.

5. Alghamdi, M.A.; Alnafei, S.H.; Radenovic, S.; Shahzad, N. Fixed point theorems for convex contraction mappings on cone metric spaces. *Math. Comput. Model.* **2011**, *54*, 2020–2026. [CrossRef]

6. Ghorbanian, V.; Rezapour, S.; Shahzad, N. Some ordered fixed point results and the property (*P*). *Comput. Math. Appl.* **2012**, *63*, 1361–1368. [CrossRef]

7. Latif, A.; Sintunavarat, W.; Ninsri, A. Approximate fixed point theorems for partial generalized convex contraction mappings in α-complete metric spaces. *Taiwan J. Math.* **2015**, *19*, 315–333. [CrossRef]

8. Miandarag, M.A.; Postolache, M.; Rezapour, S. Approximate fixed points of generalizes convex contractions. *Fixed Point Theory Appl.* **2013**, *255*, 1–8.

9. Miculescu, R.; Mihail, A. A generalization of Istrățescu's fixed piont theorem for convex contractions. *arXiv* **2015**, arXiv:1512.05490v1.

10. Khan, M.S.; Singh, Y.M.; Maniu, G.; Postolache, M. On generalized convex contractions of type-2 in b-metric and 2-metric spaces. *J. Nonlinear Sci. Appl.* **2017**, *10*, 2902–2913. [CrossRef]

11. Khan, M.S.; Singh, Y.M.; Maniu, G.; Postolache, M. On (α, p)-convex contraction and asymptotic regularity. *J. Math. Comput. Sci.* **2018**, *18*, 132–145. [CrossRef]

12. Wardowski, D. Fixed points of a new type of contractive mappings in complete metric spaces. *Fixed Point Theory Appl.* **2012**, *94*, 1–11. [CrossRef]

13. Samet, B.; Vetro, C.; Vetro, P. Fixed point theorems for α-ψ-contractive type mappings. *Nonlinear Anal.* **2012**, *75*, 2154–2165. [CrossRef]

14. Karapinar, E.; Kumam, P.; Salimi, P. On α-ψ-Meir-Keeler contractive mappings. *Fixed Point Theory Appl.* **2013**, *94*, 1–12. [CrossRef]

15. Karapinar, E. α-ψ-Greaghty contraction type mappings and some related fixed point results. *Filomat* **2014**, *28*, 37–48. [CrossRef]

16. Alsulami, H.H.; Chandok, S.; Taoudi, M.A.; Erhan, I.M. Some fixed point theorems for (α, ψ)-rational type contractive mappings. *Fixed Point Theory Appl.* **2015**, *97*, 1–12. [CrossRef]

MDPI

St. Alban-Anlage 66

4052 Basel

Switzerland

Tel. +41 61 683 77 34

Fax +41 61 302 89 18

www.mdpi.com

Mathematics Editorial Office

E-mail: mathematics@mdpi.com

www.mdpi.com/journal/mathematics

www.ingramcontent.com/pod-product-compliance
Lightning Source LLC
Chambersburg PA
CBHW051910210326
41597CB00033B/6101